Low Cholesterol Leads to an Early Death

Evidence From 101 Scientific Papers

David Evans

Foreword by
Tom Naughton

Grosvenor House
Publishing Limited

All rights reserved
Copyright © David Evans, 2012

David Evans is hereby identified as author of this
work in accordance with Section 77 of the Copyright, Designs
and Patents Act 1988

The book cover picture is copyright to David Evans

This book is published by
Grosvenor House Publishing Ltd
28-30 High Street, Guildford, Surrey, GU1 3EL.
www.grosvenorhousepublishing.co.uk

This book is sold subject to the conditions that it shall not, by way of
trade or otherwise, be lent, resold, hired out or otherwise circulated
without the author's or publisher's prior consent in any form of binding or
cover other than that in which it is published and
without a similar condition including this condition being imposed
on the subsequent purchaser.

A CIP record for this book
is available from the British Library

ISBN 978-1-78148-781-5

*This book is dedicated to my wife Julie
for her patience and support*

Also by David Evans

Cholesterol and Saturated Fat Prevent Heart Disease: Evidence From 101 Scientific Papers

Medical/liability disclaimer

This book is intended solely for informational and educational purposes and not as medical advice, nor to replace the advice of a doctor or other health-care professional. Anyone wishing to embark on any dietary or lifestyle change must first consult their health-care professional.

The decision to use any information in this book is entirely the decision of the reader, who assumes full responsibility for any and all consequences arising from such a decision. Neither the author nor the publisher shall be held liable for any consequences resulting or allegedly resulting from use of information in this book.

About the author

David is a qualified nutritional adviser. He runs a website called *Healthy Diets and Science* (www.dietsandscience.com) which is devoted to examining the scientific evidence regarding the effects of diet, pharmaceutical drugs & lifestyle on health. David has previously written a book examining the influence of cholesterol and saturated fat on heart disease. He is married to Julie, has four children, two step-children ages 14 to 28 and four grandchildren.

Table of contents

Foreword by Tom Naughton		1
Introduction		9
How to use this book		15
Chapter 1	The lower your cholesterol, the earlier you die	17
Chapter 2	High levels of both "good" and "bad" cholesterol help you to live longer	43
Chapter 3	High cholesterol does not cause cardiovascular disease	58
Chapter 4	Low cholesterol leads to an early death in many diseases	78
Chapter 5	Low cholesterol leads to an increased prevalence of many diseases	101
Chapter 6	The effects of drugs and diet	130
Chapter 7	How does cholesterol help us to live longer?	142
Chapter 8	Summary of the evidence	147
Appendix 1	Glossary	149
Appendix 2	Further resources	171
Appendix 3	List of studies	173
Index		181

Foreword

by
Tom Naughton
Health writer, researcher and filmmaker

My first job after college was as a writer and editor for a small health magazine. This was in the 1980s, when the low-fat diet craze was sweeping the country after receiving a major push from the United States Department of Agriculture (USDA) and its new Dietary Guidelines for Americans. Like most other health journalists at the time, I dutifully wrote articles warning readers to lower their cholesterol levels by cutting back on saturated fat and dietary cholesterol to avoid heart disease and other horrors. Although I'm a bit of a sceptic by nature, I'm sorry to say it never occurred to me to question whether the sudden hysteria over "killer cholesterol" (as *TIME* magazine labelled it) was based on solid scientific evidence. I just assumed it was – after all, how could the USDA, the American Heart Association, the National Institutes of Health, the Cholesterol Education Program and the revered editors of *TIME* magazine all be wrong? The very idea seemed absurd.

So I not only passed along the advice to lower cholesterol levels by avoiding saturated fat and cholesterol, I followed it, as well. I started eating cholesterol-free Egg Beaters (egg whites with added flavours and thickeners) or Grape Nuts cereal with skim milk for breakfast. I spread corn-oil margarine on my whole-wheat toast instead of butter. I sprayed butter-flavoured artificial toppings on my vegetables and microwave popcorn. I stirred fat-free non-dairy creamers into my coffee. By the time I was in my early 30s, I was living on a mostly vegetarian diet, never buying meat or real eggs at the grocery store and limiting myself to chicken or seafood when I ate out. My evening meals at home were usually based on rice, potatoes or pasta, just like the USDA recommended.

Of course, I felt quite virtuous eating this way ... which in retrospect doesn't say much about my ability to connect the dots, since my growing pride in my "healthy" diet was neatly matched by a shrinking energy level and growing waistline. But along the path to wearing size-38 pants, I picked up more than just extra weight. I also developed asthma, arthritis, psoriasis, numerous respiratory infections, occasional gastric reflux and frequent bellyaches. I was a travelling stand-up comedian by this time, and I always made sure to pack some Pepto-Bismol tablets before hitting the road. But hey, at least my cholesterol levels were "normal."

At the time, I assumed getting a little fatter and sicker as each year passed was just part of the aging process (did I mention I was only in my 30s?). It couldn't possibly be my diet to blame, since the recognised health authorities were all recommending pretty much exactly the diet I was already following.

It wasn't until I began writing and researching my documentary *Fat Head* that I began to seriously question the anti-fat, anti-cholesterol advice that all the experts (or so it seemed) had been promoting for nearly three decades. My newly rediscovered scepticism was ignited by two factors. The first was the experience that comes with age. In the twenty-some additional years of living since leaving my job at the health magazine, I'd seen the supposed experts in various scientific disciplines turn out to be wrong several times. I no longer assumed an impressive title rendered the bearer of it infallible. I'd also become increasingly aware that researchers – people I'd once naively assumed to be objective seekers of scientific truth – are often just as agenda-driven as the politicians and corporate executives who directly or indirectly pay their salaries. After all, should anyone be surprised to learn that the USDA – whose mission is to sell grains – is eager to support researchers who conclude that grains are health food?

The second and perhaps more important factor feeding my scepticism was the internet. When I was writing for a health magazine in my 20s, our "research" consisted largely of receiving press releases from the organisations I later dubbed The Usual Suspects: The American Heart Association, the USDA, the American Diabetes Association, the Centre for Science in the Public Interest, and the National Institutes of Health. We also kept large files of health articles clipped from major newspapers and magazines – most of which got their information from The Usual Suspects. To obtain those all-important authoritative quotes for our articles, we'd call doctors – who would regurgitate what they'd been told by The Usual Suspects. In other words, we were all getting our information from the same little group of gatekeepers.

In cyberspace, by contrast, the information gatekeepers are nearly non-existent. Sure, The Usual Suspects all have a major presence on the internet and still wield considerable influence, but they're not the only game in town anymore, not by a long shot. Anyone can put up a website or a blog, and sometimes it seems that everyone has. For a researcher, the astounding breadth of information available online is a gold mine ... well, at least if you don't mind picking through rather a lot of garbage to find the gold nuggets.

Once I began digging into the science of diet and health while writing the script for *Fat Head*, I quickly discovered that not everyone was on board with the anti-fat, anti-cholesterol agenda that steamrolled its way into our national consciousness in the 1980s. There were, in fact, quite a few doctors and researchers who had always vehemently disagreed with The Usual Suspects. When the internet replaced the local library as the go-to source for information, those same doctors and researchers began putting up websites and blogs that included links to published research to bolster their arguments. In a library, you'd probably never hear from those doctors and researchers unless you went looking for their work. On the internet, you're likely to trip over them by accident. I certainly did.

So there I was, attempting to write a script for a documentary about diet and health, but finding that much of the standard advice promoted by people with impressive credentials was hotly disputed by people with equally impressive credentials. Both sides cited published studies as evidence. Obviously, comparing credentials or the number of citations wasn't going to provide any clarity, so I finally did what I should have done while writing for a health magazine 25 years earlier: I taught myself to think critically about science.

I could go on and on about the scientific method and how to distinguish good studies from bad studies (and I have, on my blog and in my speeches), but here's all you need to grasp in order to understand why books like the one you're holding in your hands are so valuable: for a hypothesis to be considered scientifically validated, the evidence supporting it must be *consistent and repeatable.*

If I conduct an observational study and my data shows that people with blond hair have a higher rate of disease, that's an interesting finding. I could reasonably propose a hypothesis that blond hair is somehow related to heart disease. But if other studies show that people with dark hair have a higher rate of heart disease, while still other studies find no correlation whatsoever between hair colour and heart disease, an honest scientist could only conclude that hair colour has nothing to do with heart disease. Only a bad scientist (and there are far more of those than we'd like to believe) would engage in cherry-picking and cite my study as proof that blond hair is a risk factor for heart disease, ignoring all the contrary evidence. Then he'd sign a fat consulting contract with a pharmaceutical company to promote hair dye as a preventative therapy for heart disease.

To cite a real-world example of why consistent evidence in science matters, some years ago there was a big scare going around that power lines cause cancer. The scare began when an observational study showed that people who live near power lines have higher rates of cancer, which led to the press referring to "cancer clusters" around power lines. Cool-headed scientists tried to explain that a mere correlation didn't prove anything ... it could simply be that poor people are more likely to live near power lines, and poor people have higher rates of cancer for a number of reasons. But the press had a hot story on its hands, and bad scientists helped to feed the hysteria.

The power-line scare was finally put to rest because of a glaring inconsistency in the evidence. A large observational study of power-line workers – people exposed daily to far more of the supposedly dangerous electro-magnetic energy than anyone merely living near the power lines – showed that those workers had lower rates of cancer than the population as a whole. No consistency means no scientific validity, period – end of story.

That's why this book matters. The hypothesis that high cholesterol levels, saturated fat and dietary cholesterol are bad for your health in general and cause heart disease in particular was based on cherry-picked evidence from the beginning. If consistency is the test of scientific validity (and it always should be), the so-called diet-heart hypothesis not only doesn't pass the test, it flunks miserably. When the scientific evidence is evaluated in its entirety, it suggests that low cholesterol is more likely to shorten your life than high cholesterol.

You could spend hours and hours on the internet chasing down studies that dispute the common belief that high cholesterol levels, saturated fat and cholesterol will kill you (as I did), but thanks to David Evans, you don't have to. He's already done that work for you. In this slim but important collection of synopses from 101 studies, Evans presents the contrary evidence that the cherry pickers who still promote anti-fat, anti-cholesterol hysteria would prefer you never see. I urge you to examine this evidence for yourself. Read, think, and learn.

Then go enjoy a big meal of bacon and eggs.

Tom Naughton

Introduction

The title of this book is *not* a misprint. A thorough review of the scientific literature leads to only one conclusion – low cholesterol leads to an early death.

The above statement seems akin to a ridiculous plot line in a soap opera. However, this book is based almost entirely on the findings of peer-reviewed scientific papers. These papers show that low cholesterol levels are detrimental to your health.

Nonetheless, on an almost daily basis, we are bombarded with messages that high cholesterol is bad for our health and we must either change our diet or take a cholesterol-lowering drug to save ourselves from an early demise. These messages spring from health-care practitioners, media advertisements, friends and family.

However, the scientific literature shows the opposite, and the statement "high cholesterol is bad for your health" is a myth that started with badly designed and flawed scientific studies in the 1950s.

Many myths accepted as fact sometimes endure for centuries:

- o Up until the 1500s, some people were still executed for saying the earth was round.

- Water-filled canals were first "discovered" in 1877 on the planet Mars by Italian astronomer Giovanni Schiaparelli and were only proven to be a myth (the canals were an optical illusion caused by streaks of dust blown across the Martian surface by heavy winds) in the 1960s, when the first unmanned spacecraft made flybys over Mars and took pictures of its surface.
- Thalidomide was a sedative drug introduced in the late 1950s that was used to treat morning sickness. It was thought to be safe and was sold for four years, when it was withdrawn after being found to be a cause of birth defects.

If a myth is repeated often enough, it is eventually believed by all and sundry as true fact. As the above examples show, some myths survive for a few years, a few decades or even a few centuries.

The "high cholesterol is bad for your health" myth has survived for five or six decades.

This myth is why health care practitioners, media advertisements and friends and family keep pressing home the message we should "lower our cholesterol".

This myth is why, despite mounting scientific evidence showing the opposite, we are still advised to "lower our cholesterol".

This myth is why we are told we should eat tasteless manufactured low-fat products to "lower our cholesterol".

LOW CHOLESTEROL LEADS TO AN EARLY DEATH

This myth is why millions of healthy people are subjected to statin drugs (and their many side-effects) that will "lower our cholesterol".

This book answers the following questions:

- Does lowering our cholesterol help us to live longer?
- Does high cholesterol cause heart disease?
- Does lowering cholesterol help us to be healthier while we are living?
- What are the effects of drugs and diet on cholesterol levels and health?
- What role does cholesterol play in the maintenance of good health?

Many studies are listed that include participants aged from 20-98. These studies show unequivocally that lower cholesterol levels lead to an early death.

The effects of "good" cholesterol and "bad" cholesterol are analysed, and it is found that low levels of both "good" and "bad" are associated with higher death rates.

The role of high cholesterol in cardiovascular diseases is investigated. The result? High cholesterol does not cause heart disease.

Numerous studies are examined to find the link between cholesterol levels and different causes of death and different types of disease. This examination shows that low cholesterol levels are a common theme in many causes of death and many types of disease.

Finally, the reasons why cholesterol promotes a myriad of health benefits are evaluated.

I was inspired to research the effects of cholesterol on health after some friends and family members were advised by their doctors to take dietary action and statin medications to lower their cholesterol levels. Unfortunately, these cholesterol-lowering regimes had disastrous health consequences. These have ranged from general tiredness and debilitating "Parkinson's"-like symptoms to the early onset of cancer and premature death.

In my research, I have read many thousands of peer-reviewed scientific papers and dozens of books concerning cholesterol and health. This book is the result of that research.

The evidence shows that cholesterol is probably the most important molecule in your body and is vital for good health and essential for life. However, for commercial reasons, cholesterol is vilified. A drip feed of propaganda decrees that high cholesterol levels are destructive to our bodies, whereas low cholesterol levels are touted as the elixir that promotes good health and a long life.

Unfortunately, there are health ramifications for a public caught up in crossfire of the "high cholesterol is bad for your health" myth. To lower their cholesterol, they have to endure bland, tasteless food products and their body has to endure the toxic effects of cholesterol-lowering medications. They then suffer from more diseases and die at an earlier age.

LOW CHOLESTEROL LEADS TO AN EARLY DEATH

This book takes you through a journey that shows how low cholesterol levels lead to bad health and an earlier death, whereas higher cholesterol levels help you to be healthier and live a longer life.

The next time you are given the message to "lower your cholesterol" by health care practitioners, friends or family, show them this book with its 101 peer-reviewed scientific studies that show low cholesterol levels lead to an early death.

How to use this book

This book contains 101 scientific papers that have virtually all been published in peer-reviewed scientific journals.

The format of the book has been designed so you can analyse each paper independently, although you could read the whole book through, or pick a chapter where you need particular knowledge.

Every paper has a heading that gives the essence of its findings. Each paper also contains the name of the author, the title of the paper and where it was published.

I then try and describe the findings of the paper in a concise, easy to read manner. However some unfamiliar words are in the text, which are explained in the glossary.

For an example of the layout, Paper 8 is shown below:

Paper 8
Older men with lower cholesterol have a 45% higher death rate
Rudman, D et al. "Prognostic significance of serum cholesterol in nursing home men". *Journal of Parenteral and Enteral Nutrition.* 1988 Mar-Apr;12(2):155-8

The paper heading is "Older men with lower cholesterol have a 45% higher death rate".

D Rudman was the author. (et al means "with others"), so other people also contributed to the paper.

The title of the paper is "Prognostic significance of serum cholesterol in nursing home men".

It was published in the *Journal of Parenteral and Enteral Nutrition.*

The date, volume and page numbers of this paper are 1988 Mar-Apr;12(2):155-8

You may be wondering what peer-reviewed scientific journals are.

Peer review is a process that journals use to ensure the articles they publish represent the best scholarship currently available. When an article is submitted to a peer-reviewed journal, the editors send it out to other scholars in the same field (the author's peers) to get their opinion on the quality of the scholarship, its relevance to the field, its appropriateness for the journal, etc. If these "peers" find the proposed article does not meet these rigorous standards, then the article is rejected. This process is in stark contrast to the articles written in newspapers, which are based on circulation figures and personal opinions.

CHAPTER 1

The lower your cholesterol, the earlier you die

There is a constant barrage of information purporting the desirability of lowering our cholesterol. It's almost taken for granted that lower cholesterol must be beneficial to our health.

Indeed, many food products have been developed to actually reduce cholesterol levels. For instance, Benecol have a range of spreads and yogurts that they claim is *"proven to reduce cholesterol"*, and the Flora pro.active slogan for its mini-drinks and spread products is *"actively lowers cholesterol"*.

These products contain ingredients called plant stanol esters and plant sterols that lower cholesterol levels by blocking your body from absorbing cholesterol.

But is it wise to lower your cholesterol levels?

Will you live longer if your cholesterol levels are reduced?

The following 27 papers investigate the effects of low cholesterol levels on the death rates of men and women, ranging from young adults to those in their tenth decade.

Paper 1
Low cholesterol is associated with increased death rates in 35-74 year olds

Harris, T et al. "The low cholesterol-mortality association in a national cohort". *Journal of Clinical Epidemiology*. 1992 Jun;45(6):595-601

The research interests of the senior investigator in this study, Dr Tamara Harris, are disease and mortality in older people. In this 14-year study, she and her team examined the relationship between low cholesterol and death rates in 10,295 people (5,833 women and 4,462 men) aged 35-74.

After 14 years, Dr Harris found:

- Women whose cholesterol was below 4.1 mmol/L (158 mg/dL) had a 70% increased risk of death in comparison with women whose cholesterol was up to 5.1 mmol/L (197 mg/dL).
- Men whose cholesterol was below 4.1 mmol/L (158 mg/dL) had a 40% increased risk of death in comparison with men whose cholesterol was up to 5.1 mmol/L (197 mg/dL).

The results of the study demonstrate that low cholesterol is associated with an earlier death in both men and women, in middle age and in old age.

Paper 2
Study of 482,472 men aged 30-65 shows that low cholesterol increases the mortality rate by 35% in men

Yun-Mi, Song et al. "Which Cholesterol Level Is Related to the Lowest Mortality in a Population

with Low Mean Cholesterol Level: A 6.4-Year
Follow-up Study of 482,472 Korean Men".
American Journal of Epidemiology. Volume 151
Number 8 739-747 April 15, 2000

This large study evaluated the association between low cholesterol levels and mortality rates. The study followed 482,472 Korean men aged 30-65 years for six years.

The study found that men with the lowest cholesterol (under 135 mg/dL or 3.5 mmol/L) had a 35% increased mortality rate compared to the men with the highest cholesterol (over 252 mg/dL or 6.5 mmol/L).

Paper 3
High cholesterol results in a 13% lower death rate compared to low cholesterol in 35-64 year olds

Chen, Z et al. "Serum cholesterol concentration and coronary heart disease in population with low cholesterol concentrations". *British Medical Journal.* 1991 Aug 3;303(6797):276-82

The objective of the study, published in the prestigious *British Medical Journal*, was to investigate the connection between cholesterol levels and death rates. The study included 9,021 men and women aged 35 to 64 who were followed for up to 13 years.

The study found that those with the highest cholesterol had 13% lower death rates than those with the lowest cholesterol.

Paper 4
Higher cholesterol levels associated with lower death rates in 350,977 men

Neaton, JD et al. "Serum cholesterol level and mortality findings for men screened in the Multiple Risk Factor Intervention Trial. Multiple Risk Factor Intervention Trial Research Group". *Archives of Internal Medicine.* 1992 Jul;152(7):1490-500.

This study, headed by Professor James Neaton from the University of Minnesota, explored the link between cholesterol levels and death rates. The study included 350,977 men aged 35 to 57 who were followed for 12 years.

Professor Neaton found those men who had cholesterol levels between 200-239 mg/dL (5.1-6.2 mmol/L) had a 12% lower death rate than those men with cholesterol levels below 160 mg/dL (4.1 mmol/L).

Paper 5
In people aged 20 and over, low cholesterol is correlated with an increase in death rates

Fagot-Campagna, A et al. "Serum cholesterol and mortality rates in a Native American population with low cholesterol concentrations: a U-shaped association". *Circulation.* 1997 Sep 2;96(5):1408-15

In this study, Dr Anne Fagot-Campagna and her researchers set out to find the correlation between cholesterol levels and death rates. The study included 4,553 Pima Indians aged 20 or over who were followed for more than 12 years.

After analysing 12 years of data, the researchers found that those with the lowest cholesterol had a 9% increase in death rates compared to those with the highest cholesterol.

Paper 6
Men who stop smoking *and* lower their cholesterol levels have a 2% increase in their death rates

Multiple Risk Factor Intervention Trial Research Group (No authors listed). "Multiple Risk Factor Intervention Trial. Risk Factor Changes and Mortality Results". *Journal of the American Medical Association*. 1982;248(12):1465-1477

The trial was designed to test the effect of various health intervention initiatives on mortality rates in 12,866 high-risk men aged 35 to 57 years.

Men were assigned either:

- To a special intervention program consisting of treatment for high blood pressure, counselling for cigarette smoking and dietary advice for lowering blood cholesterol levels
- To their usual sources of health care in the community.

The study reveals that the men in the special intervention program (who quit smoking and lowered their cholesterol levels etc) had a 2% higher mortality rate than the men who did not have any health interventions.

Paper 7
15-year study of 20-95 year olds shows that low cholesterol is related to a higher death rate

Ulmer, H et al. "Why Eve is not Adam: prospective follow-up in 149,650 women and men of cholesterol and other risk factors related to cardiovascular and all-cause mortality". *Journal of Women's Health.* 2004 Jan-Feb;13(1):41-53

This study, based at the University of Innsbruck in Austria, examined the relationship between cholesterol levels and death rates in 67,413 men and 82,237 women aged 20-95 years over a 15-year period.

The study found:

- In men, a low cholesterol level is predictive of a higher risk of death at all ages.
- In women, a low cholesterol level is significantly predictive of a higher risk of death over the age of 50.

This study shows that low cholesterol is related to an increased death rate.

Paper 8
Older men with lower cholesterol have a 45% higher death rate

Rudman, D et al. "Prognostic significance of serum cholesterol in nursing home men". *Journal of Parenteral and Enteral Nutrition.* 1988 Mar-Apr;12(2):155-8

Dr Daniel Rudman, a professor of medicine at the Medical College of Wisconsin, analysed the association of cholesterol levels and death rates in 129 men (average age 70.6; range 41-96).

The analysis found:

- Men with cholesterol less than 150 mg/dL (3.9 mmol/L) had a death rate of 63% during the 14 months after the cholesterol analysis, compared to a death rate of 9% in men with cholesterol greater than 150 mg/dL (3.9 mmol/L).
- The death rate during the year after the analysis was 52% if cholesterol was below 160 mg/dL (4.1 mmol/L), compared to 7% if it was above this threshold.

The results of Dr Rudman's study show that lower cholesterol levels are associated with higher death rates in older men.

Paper 9
Men and women with higher cholesterol live longer

Isles, CJ et al. "Plasma cholesterol, coronary heart disease, and cancer in the Renfrew and Paisley survey". *British Medical Journal*. 1989;298:920-924 (8 April)

The correlation between cholesterol levels and mortality was studied by Dr Christopher Isles and his team in the general population living in the west of Scotland. The study included 7,000 men and 8,262 women aged 45-64 who were followed up for an average of 12 years.

After 12 years, Dr Isles found:

- Men with the highest cholesterol, above 254 mg/dL (6.6 mmol/L) had 10% lower death rates than men

with the lowest cholesterol, below 195 mg/dL (5.0 mmol/L).
- Women with the highest cholesterol, above 280 mg/dL (7.2 mmo/L) had 5% lower death rates than women with the lowest cholesterol, below 212 mg/dL (5.5 mmol/L).

The data from this UK study shows that those with higher cholesterol live longer.

Paper 10
18-year study finds that low cholesterol equates with higher death rates in men

Stemmermann, GN et al. "Serum cholesterol and mortality among Japanese-American men. The Honolulu (Hawaii) Heart Program". *Archives of Internal Medicine.* 1991 May;151(5):969-72

This 18-year study evaluated the association of cholesterol levels with death rates. The study included 7,961 men who were aged between 46 to 65 at the start of the study.

The study found that men with cholesterol levels below 4.65 mmol/L (180 mg/dL) had a 9.6% increase in death rates compared to men with cholesterol levels between 6.20-6.69 mmol/L (240-269 mg/dL).

Paper 11
Low cholesterol contributes to men dying at 42 times the rate of men with higher cholesterol

Rudman, D et al. "Antecedents of death in the men of a Veterans Administration nursing home". *Journal of the American Geriatric Society.* 1987 Jun;35(6):496-502

LOW CHOLESTEROL LEADS TO AN EARLY DEATH

This study, published in the distinguished *Journal of the American Geriatric Society*, investigated the correlation of different mortality predictors on death rates of 176 male residents living in a nursing home.

The study found that men with cholesterol less than or equal to 156 mg/dL (4.0 mmol/L) and hematocrit less than or equal to 41% died at a rate 42 times the rate of men with values above both thresholds.

Hematocrit refers to the percentage of the blood volume occupied by red blood cells. A low hematocrit count in the body is indicative of anaemia or haemorrhage.

There are many reasons behind anaemia. The most common type is iron-deficiency anaemia. Hematocrit levels can be increased through natural means such as intake of certain foods like meat. Red meat and liver are particularly good sources of iron. Iron from this food source is far better absorbed than plant-based sources of iron.

Paper 12
Low cholesterol increases the risk of death by at least 340% in elderly women
Forette, B et al. "Cholesterol as risk factor for mortality in elderly women". *The Lancet.* 1989 Apr 22;1(8643):868-70

This five-year study, conducted by Dr. Bernard Forette and a team of researchers from Paris, set out to determine the connection between cholesterol levels and death rates in 92 women with an average age of 82.

The study found:

- Those women with cholesterol levels of 4.0 mmol/L (154 mg/dL) had a 420% increased risk of death compared with those women with cholesterol levels of 7.0 mmol/l (270 mg/dL).
- Those women with cholesterol levels of 4.0 mmol/L (154 mg/dL) had a 340% increased risk of death compared with those women with cholesterol levels of 8.8 mmol/L (340 mg/dL).

The results from Dr Forette's study reveal that the lower your cholesterol, the earlier you die.

Paper 13
Death rates increase by 18% for every 1 mmol/L (38 mg/dL) decrease in cholesterol levels

Tuikkala, P et al. "Serum total cholesterol levels and all-cause mortality in a home-dwelling elderly population: a six-year follow-up". *Scandinavian Journal of Primary Health Care.* 2010 Jun;28(2):121-7

The study, which was based at the University of Kuopio in Finland, assessed the link between cholesterol levels and death rates in 490 elderly individuals aged 75 years or over, with a six-year follow-up.

After assessing the data, the researchers found:

- Those with cholesterol levels below 5 mmol/L (193 mg/dL) had a 52% increase in death rates compared to those with cholesterol above 6 mmol/L (232 mg/dl).
- Death rates increased by 18% for every 1 mmol/L (38 mg/dL) decrease in cholesterol levels.

The findings of the study show that people with low cholesterol have a lower survival rate than people with an elevated cholesterol level, irrespective of disease or health status.

Paper 14
Italian study shows that low cholesterol levels lead to a 47% increased mortality rate compared to high cholesterol levels

Brescianini, S et al. "Low total cholesterol and increased risk of dying: are low levels clinical warning signs in the elderly? Results from the Italian Longitudinal Study on Aging". *Journal of the American Geriatrics Society*. 2003 Jul;51(7):991-6

The objective of the study, authored by Dr Sonia Brescianini, was to analyse the relationship between cholesterol levels and all-cause mortality. The study included 4,521 men and women aged 65-84, with almost three years of follow up.

Dr Brescianini found that those with the lowest cholesterol levels had an increased death rate of 47% compared to those with the highest cholesterol levels.

Paper 15
Older men and women with low cholesterol have a 39% higher risk of death

Hu, P et al. "Does inflammation or undernutrition explain the low cholesterol-mortality association in high-functioning older persons? MacArthur studies of successful aging". *Journal of the American Geriatrics Society*. 2003 Jan;51(1):80-4

This UCLA School of Medicine study set out to determine the association between cholesterol levels and 7-year all-cause mortality in 870 older men and women.

The data from the study indicated that those men and women with lower cholesterol levels had a 39% higher risk of dying over the 7-year study than those with higher cholesterol levels.

Paper 16
20 year study shows those with the lowest cholesterol levels have a 35% increase in death rates compared to those with the highest cholesterol

Schatz, IJ et al. "Cholesterol and all-cause mortality in elderly people from the Honolulu Heart Program: a cohort study". *The Lancet.* 2001 Aug 4;358(9279):351-5

Professor Irwin Schatz and his team from the University of Hawaii, John A. Burns School of Medicine explored the relationship between cholesterol levels in 3,572 Japanese/American men (aged 71-93 years) and death rates over 20 years.

The study found that those with the lowest cholesterol had a 35% increase in mortality compared to those with the highest cholesterol.

Paper 17
Higher cholesterol levels in both men and women are linked to a longer life

Casiglia, E et al. "Total cholesterol and mortality in the elderly". *Journal of Internal medicine.* Volume 254, Issue 4, pages 353–362, October 2003

The objective of this Italian study, led by Dr Edoardo Casiglia of the University of Padua, was to evaluate whether cholesterol levels are linked with death rates and to verify whether or not this is true for both genders. The study lasted 12 years and included a total of 3,257 subjects aged 65–95 years.

Dr Casiglia analysed the data and found:

- Women with the highest cholesterol levels had a 34% lower death rate than women with the lowest cholesterol levels.
- Men with cholesterol levels over 4.66 mmol/L (180 mg/dL) and a Body Mass Index above 25 had a 28% lower death rate than men with cholesterol levels below 4.66 mmol/L (180 mg/dL) and a Body Mass Index above 25.
- Men with cholesterol levels over 4.66 mmol/l (180 mg/dL) and a Body Mass Index above 25 had a 38% lower death rate than men with cholesterol levels below 4.66 mmol/L (180 mg/dL) and a Body Mass Index below 25.

The findings from the study indicate that higher cholesterol levels in both men and women are linked to a longer life.

Paper 18
People with higher cholesterol levels have a longer lifespan

Ives, DG et al. "Morbidity and mortality in rural community-dwelling elderly with low total serum cholesterol". *Journal of Gerontology.* 1993 May;48(3):M103-7

The objective of this two-year University of Pittsburgh study was to examine the association of cholesterol levels with death rates in the elderly. The study included 3,874 participants aged 65 to 79. Those with cholesterol levels of less than 150 mg/dL (3.9 mmol/L) were compared with those with cholesterol levels between 200-240 mg/dL (5.1-6.2 mmol/L).

After two years, the study found that 12.8% of those with low cholesterol had died, whereas only 7.3% of those with the higher cholesterol had died.

The results of this study, conducted by senior researcher Diane Ives, demonstrate that higher cholesterol levels lead to a longer lifespan.

Paper 19
Older people with the highest cholesterol live the longest

Jónsson, A et al. "Total cholesterol and mortality after age 80 years". *The Lancet.* Volume 350, Issue 9093, Pages 1778 - 1779, 13 December 1997

This 15-year-long study, based in Reykjavik, compared the cholesterol levels and death rates of 105 people aged 80 years or older.

The results of the study showed that those with the lowest cholesterol (less than 5.0mmol/L 193mg/dL) had a 6% higher death rate than those with the highest cholesterol (over 6.5mmol/L 251mg/dL)

This Icelandic study reveals that older people with the highest cholesterol live the longest.

Paper 20
Low cholesterol levels are an accurate predictor of higher mortality levels in the non-demented elderly

Schupf, N et al. "Relationship between plasma lipids and all-cause mortality in non-demented elderly". *Journal of the American Geriatrics Society.* 2005 Feb;53(2):219-26

The aim of this study, headed by Dr Nicole Schupf, an associate professor of Clinical Epidemiology at Columbia University, was to investigate the relationship between cholesterol levels and the risk of death from all causes in the non-demented elderly. The study included 2,277 non-demented elderly, aged 65 to 98, living in northern Manhattan.

The study found:

- Those with the lowest cholesterol levels were approximately twice as likely to die as those with the highest cholesterol levels.
- Overall, women had higher cholesterol levels than men and lived longer.

Dr Schupf's study shows that low cholesterol levels are an accurate predictor of higher mortality levels in the non-demented elderly.

Paper 21
High cholesterol levels lead to a longer life in very old people

Dontas, AS et al. "Survival in the oldest old: death risk factors in old and very old subjects". *Journal of Aging and Health.* 1996 May;8(2):220-37

Professor Anastasios Dontas led this Greek study that assessed which factors predict survival in the ninth decade of life. The study included 210 subjects aged 68 to 79 and 287 subjects aged over 80.

Hematocrit is a measurement of how much space in the blood is occupied by red blood cells.

The research by Professor Dontas revealed:

- In those aged 68 and over, high hematocrit levels predicted survival.
- In those aged 80 or over, high cholesterol levels, 8.0 mmol/L (309 mg/dL) or over, as compared with low cholesterol levels, 5.9 mmol/L (228 mg/dL) or under, predicted survival.

The data from the study shows that high hematocrit levels and high cholesterol levels lead to a longer life in very old people.

Foods high in iron and protein, such as liver, egg yolk and beef raise hematocrit levels.

Paper 22
An analysis of 297,574 men and women finds that high cholesterol levels are linked to a longer life
Jacobs, D et al. "Report of the Conference on Low Blood Cholesterol: Mortality Associations".
Circulation. 1992 Sep;86(3):1046-60

This paper featured an analysis of 18 studies that compared cholesterol levels with death rates. The studies lasted between 9 and 30 years and included 172,760 men and 124,814 women.

The analysis found:

- Women with the highest cholesterol levels (over 240 mg/dL or 6.2 mmol/L) had 13% lower death rates than women with the lowest cholesterol (below 160 mg/dL or 4.1 mmol/L).
- Men with the highest cholesterol levels (over 240 mg/dL or 6.2 mmol/L) had 3% lower death rates than men with the lowest cholesterol (below 160 mg/dL or 4.1 mmol/L).

The result of this analysis of 18 studies shows that higher cholesterol levels are linked to a longer life.

Paper 23
Declining cholesterol rates in people over 65 are associated with a 630% increase in death rates
Grant, MD et al. "Declining cholesterol and mortality in a sample of older nursing home residents". *Journal of the American Geriatrics Society.* 1996 Jan;44(1):31-6

Dr Mark Grant, whose research interests include the treatment and prevention of aging-related diseases, headed this study, the objective of which was to examine the association between declining cholesterol levels and death rates. The study lasted for two years and included 185 participants aged 65 and older.

The data from the study revealed:

- Cholesterol levels declined 31.1 mg/dL (0.8 mmol/L) per year in the dead, whereas cholesterol

levels only declined by 4.2 mg/dL (0.1 mmol/L) per year in survivors.
- Compared with those with no change or increase in cholesterol levels, those with declining cholesterol greater than 45 mg/dL (1.1 mmol/L) per year had a 520% increase in death rates.
- Declining cholesterol levels greater than 20% per year was accompanied by a 630% increase in death rates.

Dr Grant's findings show that declining cholesterol rates in people over 65 were associated with higher death rates.

Paper 24
30% higher death rate for men with falling cholesterol levels

Iribarren, C et al. "Low serum cholesterol and mortality. Which is the cause and which is the effect?" *Circulation.* 1995 Nov 1;92(9):2396-403

This study, led by Dr Carlos Iribarren, who specialises in the health consequences of low-blood cholesterol, investigated the link between falling cholesterol levels and death rates. The study included 5,941 men aged 45 to 68 years of age who were followed for 16 years.

Dr Iribarren found:

- The death rate was 30% higher among persons with a decline from middle levels of cholesterol (180 to 239 mg/dL or 4.6 to 6.1 mmol/L) to low levels of cholesterol (less than 180 mg/dL or 4.6 mmol/L) than in persons remaining at stable middle levels of cholesterol.

- The death rate was 8% lower among persons with an increase from middle levels of cholesterol (180 to 239 mg/dL or 4.6 to 6.1 mmol/L) to high levels of cholesterol (more than 240 mg/dL or 6.1 mmol/L) than in persons remaining at stable middle levels of cholesterol.

The findings of this study indicate that falling cholesterol levels are linked with higher death rates whilst increasing cholesterol levels are linked with lower death rates.

Paper 25
Low cholesterol levels are associated with death in patients admitted to hospital

Windler, E et al. "The prognostic value of hypocholesterolemia in hospitalized patients". *Clinical Investigator.* 1994 Dec;72(12):939-43

The study examined the relationship between cholesterol levels and death rates in patients admitted to hospital. Professor Eberhard Windler from the University Hospital Eppendorf was the senior investigator in the study which included 85,463 patients and 6,543 healthy control subjects.

This German study found:

- The average cholesterol levels of patients who died, 163 mg/dL (4.2 mmol/L), was significantly lower than that of those who survived 217.8 mg/dL (5.6 mmol/L).
- The average cholesterol of surviving patients was similar to that of the 6,543 healthy control subjects.

- The death rates of patients with cholesterol levels below 100 mg/dL (2.5 mmol/L) was about ten times higher than average.
- Patients whose cholesterol levels were under 45 mg/dL (1.1 mmol/L) did not survive.

The results of Professor Windler's study show that high cholesterol levels are associated with survival and low cholesterol levels are associated with death in patients admitted to hospital.

Paper 26
Hospital patients more likely to die with low cholesterol

Crook, MA et al. "Hypocholesterolaemia in a hospital population". *Annals of Clinical Biochemistry*. 1999 Sep;36 (Pt 5):613-6

The study examined the effects of low cholesterol in patients admitted to a hospital. 57 patients, with an average age of 53, with various conditions and diseases were identified with cholesterol levels less than 3.0 mmol/L (116 mg/dL).

This UK study found:

- 18% of the patients with cholesterol levels less than 3.0 mmol/L (116 mg/dL) died during their hospitalisation.
- 39% of the patients with cholesterol levels less than 2.0 mmol/L (77 mg/dL) died during their hospitalisation.
- 71% of the patients with cholesterol levels less than 1.5 mmol/L (58 mg/dL) died during their hospitalisation.

The findings of this study show low cholesterol is associated with higher death rates in hospital patients.

Paper 27
44% increase in death rates for hospital patients with low cholesterol levels

Onder, G et al. "Serum cholesterol levels and in-hospital mortality in the elderly". *American Journal of Medicine.* 2003 Sep;115(4):265-71

The lead investigator in this study was Dr Graziano Onder from the Catholic University of the Sacred Heart, Rome. The study explored whether cholesterol levels were associated with in-hospital mortality among elderly patients. The investigators analysed cholesterol levels in 6,984 hospitalised patients aged 65 years or older.

Dr Onder found that those with the lowest cholesterol, less than 160 mg/dL (4.1 mmol/L), had 44% increased death rates compared to those with the highest cholesterol, over 240 mg/dL (6.6 mmol/L).

The results of this Italian study indicate that among older hospitalised adults, low cholesterol levels are associated with higher death rates.

The evidence from this chapter clearly shows that it is not wise to lower your cholesterol levels, as people with low cholesterol die earlier.

The studies lasted up to 30 years and include data on 1,376,547 people.

The data from the studies show that low cholesterol leads to an increased risk of death at all ages, from young adults and the middle-aged to those entering their eighth and ninth decades, and for both men and women.

Not only do the studies reveal that low cholesterol is associated with a higher death rate, but also that falling cholesterol levels lead to a reduced lifespan and lower cholesterol levels are correlated with more deaths in people that have been hospitalised.

Statistics collected from around the globe tie in with the findings of this chapter. Figure 1 features data gathered from 86 countries that compared life expectancy with cholesterol levels found a strong association between higher cholesterol levels and a higher life expectancy.

I inputted the data into Microsoft Office Excel and the resulting diagonal trend line clearly shows that as cholesterol levels increase, then life expectancy increases.

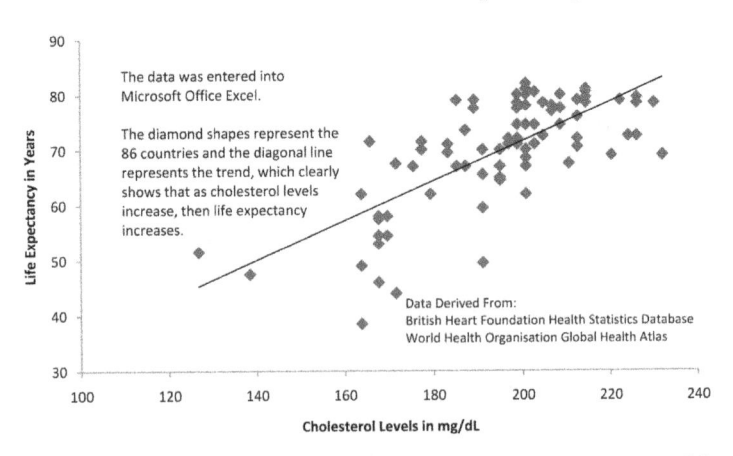

Figure 1: Data From 86 countries Comparing Life Expectancy With Cholesterol Levels

LOW CHOLESTEROL LEADS TO AN EARLY DEATH

The diagonal trend line showing increasing life expectancy at higher cholesterol levels was automatically generated by the computer programme. If you would like to test this for yourself, the data used in the graph follows.

- Column 'A' represents the country.
- Column 'B' represents the average cholesterol levels in mg/dL.
- Column 'C' represents the average cholesterol levels in mmol/L.
- Column 'D' represents the life expectancy in years.

A	B	C	D
Albania	200.85	5.19	70
Argentina	208.65	5.39	74.5
Australia	214.5	5.54	81
Austria	212.55	5.49	79
Bangladesh	163.8	4.23	62
Belarus	220.35	5.69	69
Belgium	214.5	5.54	78.5
Bolivia	195	5.04	64.5
Bosnia and Herzegovina	198.9	5.14	74.5
Brazil	196.95	5.09	71
Bulgaria	226.2	5.84	72.5
Cameroon	126.75	3.27	51.5
Canada	198.9	5.14	80
Chile	189.15	4.89	77.5
China	212.55	5.49	72
Colombia	224.25	5.79	72.5

Costa Rica	208.65	5.39	77.5
Cyprus	230.1	5.95	78.5
Czech Republic	212.55	5.49	76
Denmark	206.7	5.34	78
Dominican Republic	191.1	4.94	70
Ecuador	198.9	5.14	71.5
Egypt	187.2	4.84	67
Estonia	198.9	5.14	72
Ethiopia	169.65	4.38	54.5
Fiji	200.85	5.19	68.5
Finland	204.75	5.29	78.5
France	208.65	5.39	80
Gambia	169.65	4.38	58
Georgia	195	5.04	70
Germany	222.3	5.74	79
Ghana	167.7	4.33	57.5
Greece	185.25	4.79	79
Guatemala	185.25	4.79	67
Haiti	191.1	4.94	59.5
Hungary	204.75	5.29	72.5
Iceland	214.5	5.54	80.5
India	200.85	5.19	62
Indonesia	175.5	4.53	67
Iran	183.3	4.74	69.5
Israel	226.2	5.84	79.5
Italy	200.85	5.19	80
Jamaica	198.9	5.14	72

LOW CHOLESTEROL LEADS TO AN EARLY DEATH

Japan	200.85	5.19	82
Kenya	167.7	4.33	53
Lesotho	167.7	4.33	46
Liberia	171.6	4.43	44
Madagascar	167.7	4.33	57.5
Malawi	163.8	4.23	49
Malaysia	198.9	5.14	71.5
Malta	226.2	5.84	78.5
Mauritius	198.9	5.14	72
Mexico	187.2	4.84	73.5
Mongolia	195	5.04	65
Morocco	183.3	4.74	71
Netherlands	189.15	4.89	79
New Zealand	214.5	5.54	79.5
Nigeria	138.45	3.58	47.5
Pakistan	179.4	4.63	62
Paraguay	198.9	5.14	74.5
Peru	198.9	5.14	71
Philippines	171.6	4.43	67.5
Poland	200.85	5.19	74.5
Portugal	200.85	5.19	78
Romania	196.95	5.09	72
Russian Federation	191.1	4.94	65.5
Saint Lucia	202.8	5.24	74.5
Samoa	210.6	5.44	67.5
Saudi Arabia	177.45	4.58	70
Senegal	167.7	4.33	58

Sierra Leone	163.8	4.23	38.5
South Africa	167.7	4.33	54.5
Spain	198.9	5.14	80
Sri Lanka	212.55	5.49	70.5
Sweden	202.8	5.24	80.5
Switzerland	200.85	5.19	81
Tanzania	191.1	4.94	49.5
Thailand	202.8	5.24	71
Trinidad and Tobago	232.05	6.00	69
Tunisia	165.75	4.28	71.5
Turkey	177.45	4.58	71.5
Ukraine	200.85	5.19	67
United Arab Emirates	206.7	5.34	77
United Kingdom	198.9	5.14	78.5
United States America	198.9	5.14	77.5
Uzbekistan	195	5.04	67

This chapter may have given you another perspective from the one commonly portrayed about the effects of cholesterol levels on how long we might live.

However, there are different types of cholesterol. You may even have heard that there is "good" cholesterol and "bad" cholesterol. You may be thinking that it may be wise to at least have lower levels of the "bad" cholesterol.

The next chapter investigates the health consequences of having low or high levels of "good" or "bad" cholesterol.

Chapter 2

High levels of both "good" and "bad" cholesterol help you to live longer

A visit to the doctor for a health check-up almost invariably involves a measurement of your cholesterol levels. As well as recording your overall cholesterol level, some tests measure the individual fractions of cholesterol, such as high-density lipoprotein (HDL) cholesterol and low-density lipoprotein (LDL) cholesterol.

The different types of cholesterol play various roles. The function of LDL is to deliver cholesterol to cells, where it is used in membranes or for the synthesis of steroid hormones (testosterone, progesterone and estrogen) or to deliver fat-soluble nutrients to the tissues, whereas HDL travels in the circulation. It gathers and returns cholesterol to the liver, where it is eliminated from the body or converted into bile salts.

Despite the various types of cholesterol being used for different functions in the body, HDL cholesterol is known as the "good" cholesterol, whereas LDL cholesterol is commonly called the "bad" cholesterol.

However, do HDL and LDL cholesterol deserve their respective reputations as "good" or "bad"?

The studies in this chapter analyse the effects of HDL and LDL cholesterol on death rates.

Paper 28
53-year study shows that low levels of HDL cholesterol lead to an increased death rate
Williams, PT. "Fifty-three year follow-up of coronary heart disease versus HDL2 and other lipoproteins in Gofman's Livermore Cohort". *Journal of Lipid Research.* 2012 Feb;53(2):266-72

The investigator in this study, Paul Williams, assessed the relationship of high-density lipoprotein (HDL) cholesterol levels with total death rates and heart disease death rates. The study lasted for 53 years and included 1,905 men.

HDL cholesterol is made of HDL 2 and HDL 3. HDL 2 is larger than HDL 3.

After analysing 53 years of data, Williams found:

- Those with the lowest HDL 2 cholesterol had a 22% increase in total death rates.
- Those with the lowest HDL 2 cholesterol had a 63% increase in total heart disease death rates.
- Those with the lowest HDL 2 cholesterol had a 117% increase in premature heart disease death rates.
- Those with the lowest HDL 3 cholesterol had a 28% increase in total heart disease death rates.
- Those with the lowest HDL 3 cholesterol had a 71% increase in premature heart disease death rates.

The results of the study show that low levels of HDL cholesterol, especially HDL 2 cholesterol, are associated with higher total death rates and higher death rates from heart disease.

The best dietary way to raise HDL cholesterol levels is to eat a diet high in saturated fat *(see paper 98)*.

Paper 29
High total cholesterol levels and high levels of HDL cholesterol reduce the death rate

Volpato, S et al. "The value of serum albumin and high-density lipoprotein cholesterol in defining mortality risk in older persons with low serum cholesterol". *Journal of the American Geriatrics Society.* 2001 Sep;49(9):1142-7

The objective of the study, led by Dr Stefano Volpato from the *National Institute on Aging*, was to investigate the association between low cholesterol, albumin and death rates in older people. The study included 4,128 participants age 70 and older at (average age 78.7 years, range 70-103) who were followed for nearly five years.

Albumin is a protein in your bloodstream that helps transport a variety of important substances, including calcium, hormones, the protein bilirubin and important nutrients called fatty acids. Albumin also helps your blood maintain its osmotic pressure, which helps keep its water content from leaking through your blood vessels into surrounding tissue.

Firstly, the study found that those with low cholesterol had significantly higher death rates than those with normal and high cholesterol.

Secondly, among the participants with low cholesterol, those with albumin levels above 38 g/L had a 43% decrease in death rates compared to those with albumin levels below 38 g/L.

From those with low cholesterol and within the higher albumin group (above 38 g/L), those with levels of high-density lipoprotein (HDL) cholesterol *below* 47 mg/dL (1.2 mmol/L) had a 32% reduction in death rates compared to those with albumin below 38 g/L and HDL below 47 mg/dL (1.2 mmol/L).

From those with low cholesterol and within the higher albumin group (above 38 g/L), those with levels of high-density lipoprotein (HDL) cholesterol *above* 47 mg/dL (1.2 mmol/L) had a 62% reduction in death rates compared to those with albumin below 38 g/L and HDL below 47 mg/dL (1.2 mmol/L).

The results of the study demonstrate:

- That low cholesterol is significantly associated with higher death rates.
- In those that have low cholesterol: higher levels of albumin and HDL cholesterol are associated with lower death rates compared to those who have lower levels of albumin and HDL cholesterol.

The best way to raise your albumin levels is to eat quality protein such as beef, pork, fish, chicken and eggs.

Eating a diet rich in saturated fat raises levels of HDL cholesterol the most *(see paper 98)*.

Paper 30
Eating more saturated fat helps you live to at least 85 years of age

Rahilly-Tierney, CR et al. "Relation Between High-Density Lipoprotein Cholesterol and Survival to Age 85 Years in Men (from the VA Normative Aging Study)". *American Journal of Cardiology.* 2011 Apr 15;107(8):1173-7

This study, based at the prestigious Harvard Medical School, sought to determine whether high-density lipoprotein (HDL) cholesterol levels are associated with survival to 85 years of age. The study included 652 men (average age 65 years) who had at least 1 HDL cholesterol level measurement documented during the study and who were old enough on the date of HDL cholesterol measurement to reach 85 years of age by the end of the study.

The research revealed:

- Men who had HDL cholesterol levels higher than 50 mg/dL (1.3 mmol/L) had a 28% increased chance of survival to 85 years of age compared to men who had HDL cholesterol lower than 40 mg/dL (1.0 mmol/L).
- Each 10-mg/dL (.26 mmol/L) increase in HDL cholesterol was associated with a 14% decrease in the risk of mortality before 85 years of age.

The author of the study, Catherine Rahilly-Tierney, concluded that: *"Higher HDL cholesterol levels were significantly associated with survival to 85 years of age".*

An analysis of 27 trials found the best way to raise HDL cholesterol is to eat more saturated fat *(see paper 98)*.

Paper 31
High levels of HDL cholesterol are associated with better survival in people aged over 80

Landi, F et al. "Serum high-density lipoprotein cholesterol levels and mortality in frail, community-living elderly". *Gerontology*. 2008;54(2):71-8

This Italian study, undertaken by Dr Francesco Landi from the Catholic University in the Sacred Heart in Rome, and colleagues, evaluated the impact of high-density lipoprotein (HDL) cholesterol on death rates in older people. The two-year study analysed HDL cholesterol data from 359 subjects aged 80 years and older.

The study found:

- The HDL cholesterol level of men who died was 36.7 mg/dL (.9 mmol/L), whereas the HDL cholesterol level in the men who survived was 43.3 mg/dL (1.1 mmol/L).
- The HDL cholesterol level of women who died was 42.2 mg/dL (1.1 mmol/L), whereas the HDL cholesterol level in the women who survived was 49.3 mg/dL (1.3 mmol/L).

The results of Dr Landi's study show that high levels of HDL cholesterol are associated with better survival in people aged over 80.

Diets high in saturated fat can raise the beneficial HDL cholesterol levels *(see paper 98)*.

Paper 32
Low HDL cholesterol levels are associated with increases in deaths from heart disease and cancer

Wilson, PW et al. "High density lipoprotein cholesterol and mortality. The Framingham Heart Study". *Arteriosclerosis*. 1988 Nov-Dec;8(6):737-41

The lead investigator of the study, Professor Peter Wilson, was Director of Laboratories at the long-running Framingham Heart Study. In this 12-year study, Professor Wilson and his team examined the association of high-density lipoprotein (HDL) cholesterol levels with death rates from heart disease and cancer. The study included 2,748 participants aged 50 to 79.

The study found:

- Both men and women with the highest HDL cholesterol levels also had the highest total cholesterol levels.
- Men with the lowest HDL cholesterol levels had a 92% increase in death rates compared to the men with the highest HDL cholesterol levels.
- Women with the lowest HDL cholesterol levels had a 47% increase in death rates compared to the women with the highest HDL cholesterol levels.
- Men with the lowest HDL cholesterol levels had a 309% increase in heart disease death rates compared to the men with the highest HDL cholesterol levels.
- Women with the lowest HDL cholesterol levels had a 207% increase in heart disease death rates compared to the women with the highest HDL cholesterol levels.

- Men with the lowest HDL cholesterol levels had a 17% increase in cancer death rates compared to the men with the highest HDL cholesterol levels.
- Women with the lowest HDL cholesterol levels had an 8% increase in cancer death rates compared to the women with the highest HDL cholesterol levels.

The results of this study show that men and women with the lowest HDL cholesterol levels also had the lowest total cholesterol levels, and that low HDL cholesterol levels are associated with increases in deaths from heart disease and cancer.

The most effective way to raise HDL cholesterol levels is to consume a diet high in saturated fat *(see paper 98)*.

Paper 33
Low HDL cholesterol and low LDL cholesterol levels are linked to an earlier death

Weverling-Rijnsburger, AW et al. "High-density vs low-density lipoprotein cholesterol as the risk factor for coronary artery disease and stroke in old age". *Archives of Internal Medicine.* 2003 Jul 14;163(13):1549-54

The aim of this study was to evaluate the relationships between cholesterol levels, low-density lipoprotein (LDL) cholesterol levels and high-density lipoprotein (HDL) cholesterol levels and death rates. The study included 705 participants who had reached the age of 85 who were followed for four years.

Dr Annelies W E Weverling-Rijnsburger, from the Leiden University Medical Centre, who conducted the study, found the following:

- Those with the lowest cholesterol (163-195 mg/dL or 4.2-5.0 mmol/L) had a 60% increase in death rates compared to those with the highest cholesterol (248-280 or 6.4-7.2 mmol/L).
- Those with the lowest LDL cholesterol (94-116 mg/dL or 2.4-3.0 mmol/l) had a 40% increase in death rates compared to those with the highest LDL cholesterol (165-196 or 4.3-5.0 mmol/L).
- Those with the lowest HDL cholesterol (32-40 mg/dL or .8-1.0 mmol/L) had a 70% increase in death rates compared to those with the highest HDL cholesterol (60-73 or 1.5-1.9 mmol/L).

The findings of this study show that lower levels of cholesterol, low-density lipoprotein (LDL) cholesterol and high-density lipoprotein (HDL) cholesterol are linked to an earlier death.

Paper 34
Middle-aged men have lower death rates with higher levels of HDL and LDL cholesterol
Cullen, P et al. "The Münster Heart Study (PROCAM). Total Mortality in Middle-Aged Men Is Increased at Low Total and LDL Cholesterol Concentrations in Smokers but Not in Non-smokers".
Circulation. 1997; 96:2128-2136

This German study investigated the relationship of cholesterol levels with death rates in 10,856 men aged 36 to 65 with up to 14 years of follow up.

The study found:

- Those with cholesterol levels between 213-231 mg/dL (5.5-6.0 mmol/L) had a 9.4% decreased risk

of death compared to those with cholesterol levels below 190 mg/dL (4.9 mmol/L).
- Those with low-density lipoprotein (LDL) cholesterol levels between 138-155 mg/dL (3.5-4.0 mmol/L) had an 18% decreased risk of death compared to those with low-density lipoprotein (LDL) cholesterol levels below 117 mg/dL (3.0 mmol/L).
- Those with high-density lipoprotein (HDL) cholesterol levels over 55 mg/dL (1.4 mmol/L) had a 76% decreased risk of death compared to those with high-density lipoprotein (HDL) cholesterol levels below 37 mg/dL (.95 mmol/L).

The data from this study shows there is a decrease in death rates at higher levels of total cholesterol, LDL cholesterol and HDL cholesterol in middle-aged men compared to lower levels of total cholesterol, LDL cholesterol and HDL cholesterol.

Paper 35
Low cholesterol levels, and in particular, low levels of LDL cholesterol are associated with a shorter lifespan
Akerblom, JL et al. "Relation of plasma lipids to all-cause mortality in Caucasian, African-American and Hispanic elders". *Age and Ageing.* 2008 Mar;37(2):207-13

The objective of this study was to investigate the relationship of cholesterol levels to all-cause death rates in the non-demented elderly. The study, which was based at Columbia University, included 2,556 non-demented elderly, aged between 65-103 years, who were subject to 8,846 person years of follow-up. Among the participants, 66.1% were women, 27.6% were white, 31.2% were African-American and 41.2% were Hispanic.

The results of the study revealed:

- The whites with the lowest cholesterol had a 120% increase in death rates compared to the whites with the highest cholesterol.
- The African-Americans with the lowest cholesterol had a 90% increase in death rates compared to the African-Americans with the highest cholesterol.
- Cholesterol levels were not related to death rates in Hispanics.
- The whites with the lowest levels of low-density lipoprotein (LDL) cholesterol had a 80% increase in death rates compared to the whites with the highest levels of low-density lipoprotein (LDL) cholesterol.
- The African-Americans with the lowest levels of low-density lipoprotein (LDL) cholesterol had a 90% increase in death rates compared to the African-Americans with the highest levels of low-density lipoprotein (LDL) cholesterol.
- The Hispanics with the lowest levels of low-density lipoprotein (LDL) cholesterol had a 40% increase in death rates compared to the Hispanics with the highest levels of low-density lipoprotein (LDL) cholesterol.

This study indicates that lower cholesterol levels, and in particular, lower levels of LDL cholesterol, are associated with a shorter lifespan.

Paper 36
High levels of LDL Cholesterol lead to a longer life
Fried, LP et al. "Risk factors for 5-year mortality in older adults: the Cardiovascular Health Study". *Journal of the American Medical Association.* 1998 Feb 25;279(8):585-92

Diuretics, also known as water pills (such as Demadex, Diuril, Enduron, Esidrix, Lasix, Lozol, Saluron, Thalitone, Zaroxolyn, Oretic, Aldactone and Hydrozyne) are prescribed to lower blood pressure. Low-density lipoprotein (LDL) cholesterol is supposedly the "bad" cholesterol and people are prescribed statin drugs to lower their LDL cholesterol.

Dr Linda Fried is a Professor of Medicine, Epidemiology, Health Policy and Nursing at the Johns Hopkins Medical Institutions. In this study, Dr Fried and her researchers set out to determine the factors that predict mortality in men and women aged 65 years or older. The study included 5,201 men and women who were followed for five years.

The researchers found:

- Those who took diuretics had a 103% increase in death rates compared to those who did not take diuretics.
- Those with the highest levels of LDL cholesterol had a 49% reduction in death rates compared to those with the lowest levels of LDL cholesterol.

Dr Fried's study reveals that *not* taking diuretics and having high levels of low-density lipoprotein (LDL) cholesterol will result in a longer life.

Paper 37
Low LDL cholesterol levels are associated with an earlier death

Ogushi, Y et al. "Blood cholesterol as a good marker of health in Japan". *World Review of Nutrition and Dietetics*. 2009;100:63-70

This study examined the relationship between low-density lipoprotein (LDL) cholesterol and death rates. The study lasted for eight years and included 26,000 men and women.

After analysing eight years of data, the investigators found:

- The death rate of men whose LDL cholesterol levels were below 100 mg/dL (2.6 mmol/L) were higher than in men whose LDL cholesterol levels were between 100 mg/dL and 160 mg/dL (2.6-4.1 mmol/L).
- The death rate of women whose LDL cholesterol levels were below 120 mg/dL (2.6 mmol/L) were higher than in women whose LDL cholesterol levels were above 120 mg/dL (2.6 mmol/L).

The findings from this study show that lower LDL cholesterol levels are associated with an earlier death.

Paper 38
High levels of LDL cholesterol are associated with lower death rates and lower rates of cardiovascular disease
Tikhonoff, V et al. "Low-density lipoprotein cholesterol and mortality in older people". *Journal of the American Geriatrics Society.* 2005 Dec;53(12):2159-64

The objective of the study, led by Dr Valerie Tikhonoff from the University of Padua, Italy, was to investigate the role of low-density lipoprotein (LDL) cholesterol as a predictor of mortality in elderly subjects. The study included 3,120 subjects aged 65 and older who were followed for 12 years.

The study found:

- Men who had the highest LDL cholesterol had a 34% decrease in death rates compared to the men with the lowest LDL cholesterol.
- Women who had the highest LDL cholesterol had a 48% decrease in death rates compared to the women with the lowest LDL cholesterol.
- Men who had the highest LDL cholesterol had an 8% decrease in cardiovascular disease death rates (stroke, heart attack, heart failure) compared to the men with the lowest LDL cholesterol.
- Women who had the highest LDL cholesterol had a 23% decrease in cardiovascular disease death rates compared to the women with the lowest LDL cholesterol.

The results of this study demonstrate that high levels of LDL cholesterol are associated with lower death rates and lower death rates of cardiovascular disease.

To answer the question at the start of the chapter - do HDL and LDL cholesterol deserve their reputations as "good" and "bad" respectively – the answer is yes and no.

- Yes to HDL cholesterol having a "good" reputation. High levels of HDL were associated with up to a 92% lower death rate and up to a 309% decreased

risk of death from heart disease. The data also revealed men and women with higher HDL levels had lower rates of death from cancer.
- No to LDL cholesterol having a "bad" reputation. Similar to HDL, high levels of LDL cholesterol correlated with a lower overall death rate and lower rates of heart disease.

The findings of these studies provide hard evidence that high levels of both HDL and LDL cholesterol lead to lower death rates and also a decreased risk of dying from heart disease and cancer.

So far, we have examined the scientific literature and found that high cholesterol helps you to live longer and that high levels of both HDL and LDL cholesterol may also help you to increase your lifespan.

Nonetheless, despite these findings that high cholesterol leads to a longer life, many people are concerned that high levels of cholesterol will put them at an increased risk from cardiovascular diseases such as heart attacks, stroke and heart failure.

The next chapter trawls through the scientific literature to evaluate if these concerns are justified.

CHAPTER 3

High cholesterol does not cause cardiovascular disease

Heart disease is a type of cardiovascular disease. Cardiovascular disease refers to any disease that affects the cardiovascular system such as stroke, angina or heart failure.

With all the hype regarding cholesterol levels, the term "high cholesterol causes heart disease" seems as natural as saying fish live in water.

The "high cholesterol causes heart disease" hypothesis was initially based on very shaky and flimsy evidence from outdated and biased research conducted in the 1950s.

Enormous amounts of money have since been spent in attempting to prove the hypothesis. All attempts have failed. Despite this failure, the hypothesis is embedded in the psyche of many nations.

Concurrent to this all pervading hypothesis, the food processing and pharmaceutical industries have based their best-selling products on the "high cholesterol causes heart disease" paradigm, resulting in supermarkets been inundated with insipid low-fat, low cholesterol products and doctors monumentally overprescribing cholesterol lowering drugs.

But are high cholesterol levels responsible for cardiovascular diseases?

The following studies give telling answers to this question.

Paper 39
Higher death rates, increased stroke and heart disease associated with low cholesterol

Tilvis, RS et al. "Prognostic significance of serum cholesterol, lathosterol, and sitosterol in old age; a 17-year population study". *Annals of Medicine.* 2011 Jun;43(4):292-301

Professor Reijo Tilvis presided over this 17-year study based at the University of Helsinki. The study investigated the effects of cholesterol levels on death rates in 623 people over 75 years of age.

The 17-year study revealed:

- Total cholesterol declined in old age, and low cholesterol was associated with poor health.
- Cholesterol below 5.0 mmol/L (193 mg/dL) was associated with a 54% accelerated death rate.
- Cholesterol below 5.0 mmol/L (193 mg/dL) was associated with a 113% increase in stroke and heart disease.

Professor Tilvis concludes: *"Low cholesterol levels are associated with deteriorating health and indicate impaired survival in old age."*

Paper 40
Rates of heart disease deaths are higher with low cholesterol

Behar, S et al. "Low total cholesterol is associated with high total mortality in patients with coronary heart disease. The Bezafibrate Infarction Prevention (BIP) Study Group". *European Heart Journal.* 1 Jan 1997; 18(1): 52-9

The author of this study, Professor Solomon Behar, is a senior cardiologist at the Sheba Medical Center in Israel. The study was undertaken to explore the relationships between pre-existing low cholesterol and death rates. Cholesterol levels were measured in 11,563 patients with heart disease who were followed for 3.3 years.

The patients were classified into two groups:

- Cholesterol levels of 160mg/dL (4.1mmol/L) or under.
- Cholesterol levels over 160mg/dL (4.1mmol/L).

After dividing the patients into these two groups, Professor Behar found:

- Those with low cholesterol had 9% more heart disease deaths.
- Those with low cholesterol had a 49% higher total death rate.
- Those with low cholesterol had a 127% higher risk of non heart disease deaths.
- The most frequent cause of non heart disease death associated with low total cholesterol was cancer.

This study reveals that if you have coronary heart disease and have low cholesterol, you are about one and a half times more likely to die than if you have higher cholesterol.

Paper 41
Low cholesterol leads to 30% higher death rates from vascular causes

Räihä, I et al. "Effect of Serum Lipids, Lipoproteins, and Apolipoproteins on Vascular and Nonvascular Mortality in the Elderly". *Arteriosclerosis, Thrombosis, and Vascular Biology.* 1997;17:1224-1232

The purpose of this 11-year study was to determine the effect of cholesterol levels on mortality rates from vascular and nonvascular causes. The study included 347 individuals aged 65 years or older. Vascular causes included heart disease, heart failure, stroke etc.

The study was headed by Dr Ismo Räihä, a specialist in Internal Medicine, who found:

- Nonvascular causes of death were 80% higher for those with the lowest cholesterol (less than 5.0 mmol/l or 193 mg/dL) compared with those of the highest cholesterol (over 8.0 mmol/L or 309 mg/dL).
- Vascular causes of death were 30% higher for those with the lowest cholesterol (less than 5.0 mmol/l or 193 mg/dL) compared with those of the highest cholesterol (over 8.0 mmol/L or 309 mg/dL).

The findings of the study reveal that low cholesterol levels are associated with higher death rates from nonvascular and vascular causes.

Paper 42
A direct association between falling cholesterol levels and increased death rates from cardiovascular diseases

Anderson, KM et al. "Cholesterol and mortality.
30 years of follow-up from the Framingham study".
Journal of the American Medical Association.
1987 Apr 24;257(16):2176-80

This study analysed the connection between cholesterol levels and total death rates and cardiovascular death rates. Cholesterol levels were measured in 1,959 men and 2,415 women aged between 31 and 65 years who were free of cardiovascular disease and cancer.

After analysing 30 years of data, the researchers found:

- There is a direct association between falling cholesterol levels over the first 14 years and an increase in total death rates over the following 18 years (11% increase in deaths per 1 mg/dL (.03 mmol/L) per year drop in cholesterol levels).
- There is a direct association between falling cholesterol levels over the first 14 years and an increase in cardiovascular death rates over the following 18 years (14% increase in deaths per 1 mg/dL (.03 mmol/L) per year drop in cholesterol levels).

The study illustrates that if your cholesterol levels fall, you have a greatly increased risk of dying – and an even greater risk of dying from cardiovascular disease. It also makes clear that the more your cholesterol levels fall, the bigger the risk of premature death.

Paper 43
Higher cholesterol levels reduce the risk of cardiovascular diseases

Petursson, H et al. "Is the use of cholesterol in mortality risk algorithms in clinical guidelines valid? Ten years prospective data from the Norwegian HUNT 2 study". *Journal of Evaluation in Clinical Practice.* 2012 Feb;18(1):159-68

The study (ten years' duration) investigated whether cholesterol levels are a risk factor for mortality in 52,087 individuals (24,235 men and 27,852 women) aged 20-74 years and free from known cardiovascular disease at the start of the study.

The study found:

- Compared with women whose cholesterol was under 5.0 mmol/L (193 mg/dL), those with a reading over 7.0 mmol/L (270 mg/dL) enjoyed a 28% reduction of death.
- Compared with men whose cholesterol was under 5.0 mmol/L (193 mg/dL), those with a reading over 7.0 mmol/L (270 mg/dL) enjoyed a 11% reduction of death.
- Compared with women whose cholesterol was under 5.0 mmol/L (193 mg/dL), those with a reading over 7.0 mmol/L (270 mg/dL) enjoyed a 26% reduction of cardiovascular diseases.
- Compared with men whose cholesterol was under 5.0 mmol/L (193 mg/dL), those with a reading up to 5.9 mmol/L (228 mg/dL) enjoyed a 20% reduction of cardiovascular diseases.

The results from this Norwegian study indicate at if you have cholesterol levels above 5.0 mmol/L (193 mg/dL), you will live longer and have less cardiovascular diseases.

Paper 44
Higher cholesterol levels and higher meat consumption are associated with decreased rates of heart disease deaths

Luoma, PV et al. "High serum alpha-tocopherol, albumin, selenium and cholesterol, and low mortality from coronary heart disease in northern Finland". *Journal of Internal Medicine*. 1995 Jan;237(1):49-54

The study investigated the risk factors for, and the rates of heart disease deaths in northernmost Finland compared with southern areas of Finland. The study lasted for nine years and included 350 participants with an average age of 46 years.

The senior investigator of the study was Professor Pauli Luoma from the University of Oulu. His research uncovered that:

- The death rates from heart disease were 17% lower in northernmost Finland compared with southern areas of Finland.
- Cholesterol levels were 6.3% higher in northernmost Finland compared with southern areas of Finland.
- Low-density lipoprotein (LDL) cholesterol levels were 7.0% higher in northernmost Finland compared with southern areas of Finland.

- Vitamin E levels were 14.2% higher in northernmost Finland compared with southern areas of Finland.
- Vitamin E levels increased with the consumption of reindeer meat.

The results of the study show that higher cholesterol levels and higher meat consumption are associated with decreased rates of heart disease deaths.

Paper 45
Heart attack survivors live longer if they have high cholesterol

Foody, JM et al. "Long-term prognostic importance of total cholesterol in elderly survivors of an acute myocardial infarction: the Co-operative Cardiovascular Pilot Project". *Journal of the American Geriatrics Society.* 2003 Jul;51(7):930-6

This Yale School of Medicine study was headed by Dr Joanne Foody, who is an assistant professor of medicine. The study sought to determine the relationship of cholesterol levels to long-term survival rates in elderly survivors of a heart attack. The study analysed the death rates of 4,923 heart attack patients aged 65 and older for six years.

The study revealed:

- After one year those with the lowest cholesterol levels (less than 160 mg/dL or 4.1 mmol/L) had 5% more deaths than those with the highest cholesterol (above 240 mg/dL or 6.2 mmol/L).
- After six years those with the lowest cholesterol levels (less than 160 mg/dL or 4.1 mmol/L) had

7.6% more deaths than those with the highest cholesterol (above 240 mg/dL or 6.2 mmol/L).

The results of Dr Foody's study show that heart attack survivors live longer if they have high cholesterol.

Paper 46
In patients with chronic heart failure, higher cholesterol levels are associated with a longer survival time

Rauchhaus, M et al. "The relationship between cholesterol and survival in patients with chronic heart failure". *Journal of the American College of Cardiology*. 2003; 42:1933-1940

The objective of the study, conducted by Dr Mathias Rauchhaus and his team, was to describe the relationship between cholesterol and survival in patients with chronic heart failure. A total of 417 patients were involved in the study.

The results of the study revealed:

- Higher total cholesterol levels were a predictor of survival.
- The chance of survival increased 25% for each mmol/L (38 mg/dL) increment in cholesterol.
- After one year, those with cholesterol levels above 5.2 mmol/L (201 mg/dL) had a 17% higher survival rate than those with cholesterol levels below 5.2 mmol/L (201 mg/dL).
- After three years, those with cholesterol levels above 5.2 mmol/L (201 mg/dL) had a 22% higher

survival rate than those with cholesterol levels below 5.2 mmol/L (201 mg/dL).

The findings of the study show that in patients with chronic heart failure, higher cholesterol levels are associated with a longer survival time.

Paper 47
Death rates in stroke, heart failure, and cancer are elevated in people with low cholesterol levels

Nago, N et al. "Low cholesterol is associated with mortality from stroke, heart disease, and cancer: the Jichi Medical School Cohort Study". *Journal of Epidemiology.* 2011;21(1):67-74

The senior investigator of this study was Dr Naoki Nago from the Tokyo-kita Social Health Insurance Hospital. The study of 12,334 healthy adults aged 40 to 69 years investigated the relationship between low cholesterol and mortality and examined whether that relationship differs with respect to cause of death.

The study found:

- Men with the lowest cholesterol (under 4.14mmol/L or 160mg/dL) had a 37% higher death rate than men with the highest cholesterol (over 6.21 mg/dL or 240 mmol/L).
- Women with the lowest cholesterol (under 4.14mmol/L or 160mg/dL) had a 53% higher death rate than women with the highest cholesterol (over 6.21 mg/dL or 240 mmol/L).
- The risk of death in the lowest cholesterol group for hemorrhagic stroke, heart failure (excluding

myocardial infarction), and cancer mortality was significantly higher than those of the moderate cholesterol group, for each cause of death.

Dr Nago concluded: *"Low cholesterol was associated with increased risks of cancer, hemorrhagic stroke, and heart failure excluding myocardial infarction."*

Paper 48
Heart failure patients with low cholesterol levels have a 240% increased risk of urgent transplant and death
Afsarmanesh, N et al. "Total cholesterol levels and mortality risk in nonischemic systolic heart failure". *American Heart Journal.* 2006 Dec;152(6):1077-83

Nonischemic heart disease is a disease of the heart that lacks the associated coronary artery disease often found in other diseases of the heart. It's usually linked to a disease in one or more of the cardiac muscles, causing the heart to pump in an ineffective manner, thereby reducing the transport of blood, oxygen and other nutrients throughout the body. One of the more common nonischemic heart diseases is dilated cardiomyopathy. In this form of heart disease, your left ventricle has weakened (low left ventricular ejection fraction (LVEF)) to the point where it can no longer pump enough blood.

Hemodynamics is a measurement of blood pressure and blood flow.

The New York Heart Association (NYHA) Functional Classification provides a simple way of classifying the extent of heart failure. It places patients in one of four classes based on how much they are limited during physical activity. For instance, Class One indicates: *No symptoms and no limitation in ordinary physical activity*, whilst Class Four is defined as: *Severe limitations. Experiences symptoms even while at rest. Mostly bedbound patients.*

In this study, Dr Nasim Afsarmanesh, who is an Assistant Clinical Professor of Internal Medicine and Neurosurgery at the University of California, Los Angeles, analysed the cholesterol levels of 614 patients with nonischemic systolic heart failure who had a left ventricular ejection fraction less than 40%.

The study found:

- Patients with lower cholesterol levels had a lower left ventricular ejection fraction.
- Patients with lower cholesterol levels had worse hemodynamic profiles.
- Patients with lower cholesterol levels had a higher New York Heart Association class.
- Patients with the lowest cholesterol levels had a 240% increased risk of urgent transplant and death compared to patients with the highest cholesterol.

Dr Afsarmanesh's study demonstrates that low cholesterol levels are strongly associated with increased death rates in patients with nonischemic, systolic heart failure.

Paper 49
Higher cholesterol levels are associated with less severe strokes and lower death rates
Olsen, TS et al. "Higher total serum cholesterol levels are associated with less severe strokes and

lower all-cause mortality: ten-year follow-up of
ischemic strokes in the Copenhagen Stroke Study".
Stroke. 2007 Oct;38(10):2646-51

The Scandinavian Stroke Scale is a measurement of stroke severity where a score of 0 is the most severe stroke and a score of 58 is the least severe stroke.

This Danish-based study investigated the relationship between cholesterol levels and both stroke severity and post-stroke death rates in 513 stroke patients with an average age of 75 who were followed for ten years.

The Hvidovre University Hospital study found:

- Each 1 mmol/L (38 mg/dL) increase in cholesterol levels resulted in an increase in the Scandinavian Stroke Scale score of 32% meaning that higher cholesterol levels are associated with less severe strokes.
- Each 1 mmol/L (38 mg/dL) increase in cholesterol levels resulted in an 11% decrease in death rates.

The results of the study show that higher cholesterol levels are associated with less severe strokes and lower death rates.

Paper 50
Low cholesterol levels increase the risk of death from stroke, cancer and all-causes

Iribarren, C et al. "Serum total cholesterol and mortality. Confounding factors and risk modification in Japanese-American men". *Journal of the American Medical Association.* 1995 Jun 28;273(24):1926-32

This study was conducted by Dr Carlos Iribarren, who is an Adjunct Assistant Professor in the Department of Epidemiology and Biostatistics at the University of California. Dr Iribarren and his research team set out to determine the connection between cholesterol levels and death rates due to major causes. The study included 7,049 middle-aged men, who were followed for 23 years.

Dr Iribarren found:

- Men with the lowest cholesterol levels, below 4.66 mmol/L (180 mg/dL), had a 141% increased risk of death from hemorrhagic stroke compared to the men with cholesterol levels up to 6.19 mmol/L (239 mg/dL).
- Men with the lowest cholesterol levels, below 4.66 mmol/L (180 mg/dL), had a 41% increased risk of death from cancer compared to the men with cholesterol levels up to 6.19 mmol/L (239 mg/dL).
- Men with the lowest cholesterol levels, below 4.66 mmol/L (180 mg/dL), had a 23% increased risk of death compared to the men with cholesterol levels up to 6.19 mmol/L (239 mg/dL).

The results of this University of California study show that middle-aged men with low cholesterol are at greater risk of death from stroke, cancer and all-causes compared to men with higher cholesterol.

Paper 51
Patients hospitalised with a stroke with low cholesterol have a 117% increased risk of death compared to patients with high cholesterol

Zuliani, G et al. "Low cholesterol levels are associated with short-term mortality in older patients with ischemic stroke". *Journals of Gerontology. Series A Biological Sciencies and Medical Sciencies.* 2004 Mar;59(3):293-7

The study evaluated the association between cholesterol levels and 30-day death rates in 490 older patients admitted to hospital with ischemic stroke.

The study found that those with the lowest cholesterol levels, under 4.1 mmol/L (158 mg/dL), had a 117% increased risk of death compared with those with the highest cholesterol, over 5.2 mmol/L (201 mg/dL).

Despite the hype and hysteria to lower our cholesterol levels as much as possible to avoid succumbing to a cardiovascular disease, the studies in this chapter show a rather different outcome.

The death rates of all types of cardiovascular diseases, such as heart disease, heart attack, heart failure, hemorrhagic stroke and ischemic stroke (and also cancer) are shown to actually *decrease* at higher cholesterol levels.

The results of these studies correlate with worldwide statistics regarding cholesterol levels and deaths from cardiovascular diseases.

Figure 2 shows data from 86 countries that compares death rates from cardiovascular diseases with cholesterol levels. The data for Figure 2 was derived from the British Heart Foundation statistics database and the World Health Organisation global health atlas.

LOW CHOLESTEROL LEADS TO AN EARLY DEATH

As in Figure 1, I inputted the data into Microsoft Office Excel and the resulting diagonal trend, which clearly shows that as cholesterol levels increase, then death from cardiovascular diseases decrease.

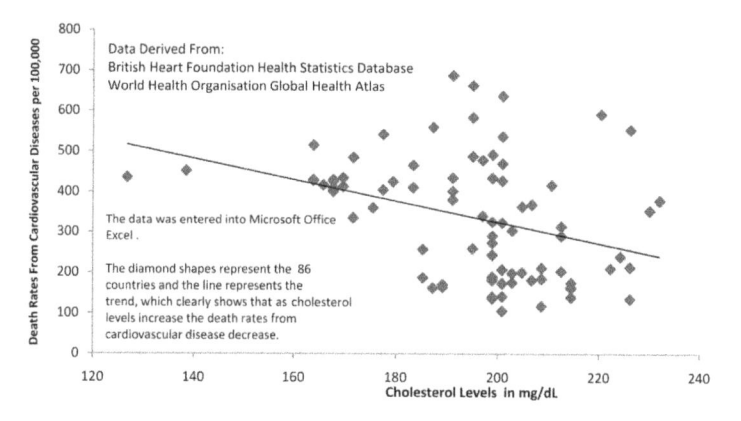

Figure 2: Data From 86 Countries Comparing Death Rates From Cardiovascular Diseases With Cholesterol Levels

The computer programme automatically generated the diagonal trend line showing the decreasing rates of heart disease at higher cholesterol levels. Again, if you would like to test this for yourself, the data used in the graph follows.

- Column 'A' represents the country.
- Column 'B' represents the average cholesterol levels in mg/dL.
- Column 'C' represents the average cholesterol levels in mmol/L.
- Column 'D' represents the death rates from cardiovascular diseases per 100,000 of the population.

A	B	C	D
Albania	200.85	5.19	537
Argentina	208.65	5.39	212
Australia	214.5	5.54	140
Austria	212.55	5.49	204
Bangladesh	163.8	4.23	428
Belarus	220.35	5.69	592
Belgium	214.5	5.54	162
Bolivia	195	5.04	260
Bosnia and Herzegovina	198.9	5.14	492
Brazil	196.95	5.09	341
Bulgaria	226.2	5.84	554
Cameroon	126.75	3.27	436
Canada	198.9	5.14	141
Chile	189.15	4.89	165
China	212.55	5.49	291
Colombia	224.25	5.79	240
Costa Rica	208.65	5.39	185
Cyprus	230.1	5.95	354
Czech Republic	212.55	5.49	315
Denmark	206.7	5.34	182
Dominican Republic	191.1	4.94	381
Ecuador	198.9	5.14	244
Egypt	187.2	4.84	560
Estonia	198.9	5.14	435
Ethiopia	169.65	4.38	435
Fiji	200.85	5.19	470

LOW CHOLESTEROL LEADS TO AN EARLY DEATH

Finland	204.75	5.29	201
France	208.65	5.39	118
Gambia	169.65	4.38	413
Georgia	195	5.04	584
Germany	222.3	5.74	211
Ghana	167.7	4.33	404
Greece	185.25	4.79	258
Guatemala	185.25	4.79	188
Haiti	191.1	4.94	402
Hungary	204.75	5.29	364
Iceland	214.5	5.54	164
India	200.85	5.19	428
Indonesia	175.5	4.53	361
Iran	183.3	4.74	466
Israel	226.2	5.84	136
Italy	200.85	5.19	174
Jamaica	198.9	5.14	326
Japan	200.85	5.19	106
Kenya	167.7	4.33	401
Lesotho	167.7	4.33	404
Liberia	171.6	4.43	485
Madagascar	167.7	4.33	430
Malawi	163.8	4.23	430
Malaysia	198.9	5.14	274
Malta	226.2	5.84	214
Mauritius	198.9	5.14	434
Mexico	187.2	4.84	163

Mongolia	195	5.04	488
Morocco	183.3	4.74	411
Netherlands	189.15	4.89	171
New Zealand	214.5	5.54	175
Nigeria	138.45	3.58	452
Pakistan	179.4	4.63	425
Paraguay	198.9	5.14	291
Peru	198.9	5.14	190
Philippines	171.6	4.43	336
Poland	200.85	5.19	324
Portugal	200.85	5.19	208
Romania	196.95	5.09	479
Russian Federation	191.1	4.94	688
Saint Lucia	202.8	5.24	304
Samoa	210.6	5.44	417
Saudi Arabia	177.45	4.58	405
Senegal	167.7	4.33	426
Sierra Leone	163.8	4.23	515
South Africa	167.7	4.33	410
Spain	198.9	5.14	137
Sri Lanka	212.55	5.49	314
Sweden	202.8	5.24	176
Switzerland	200.85	5.19	142
Tanzania	191.1	4.94	435
Thailand	202.8	5.24	199
Trinidad and Tobago	232.05	6.00	379
Tunisia	165.75	4.28	417

LOW CHOLESTEROL LEADS TO AN EARLY DEATH

Turkey	177.45	4.58	542
Ukraine	200.85	5.19	637
United Arab Emirates	206.7	5.34	369
United Kingdom	198.9	5.14	182
United States America	198.9	5.14	188
Uzbekistan	195	5.04	663

If we recap the last three chapters, the scientific evidence has revealed that low cholesterol leads to a higher death rate, that low levels of HDL and LDL cholesterol are also associated with an earlier demise, and finally, that low cholesterol levels increase the risk of dying from cardiovascular diseases.

You may be wondering in what other diseases and conditions low cholesterol makes an early death more likely.

The next chapter examines the effect of cholesterol levels on 41 types of death.

CHAPTER 4

Low cholesterol leads to an early death in many diseases

The cholesterol propaganda machine is still omnipresent in society, and a measurement of "high cholesterol" sounds like a death sentence to many ears.

The first three chapters of this book have produced scientific evidence that show that actually, "high cholesterol" is a ticket to longevity and low cholesterol is more likely to be a death sentence.

Low cholesterol levels affect the outcome of life or death in many serious illnesses. The following 23 papers look into the areas where low cholesterol levels may be a health catastrophe and death is more frequent.

Paper 52
As cholesterol levels are lowered deaths from accidents, suicides and homicides increase by up to 30%
Muldoon, MF et al. "Lowering cholesterol concentrations and mortality: a quantitative review of primary prevention trials". *British Medical Journal*. 1990 Aug 11;301(6747):309-14

This University of Pittsburgh study analysed the findings of six cholesterol reduction trials. The participants in the six cholesterol reduction trials received either diet-based, drug-based, a mixture of diet and drug cholesterol-lowering treatment or placebo. The trials lasted for an average of 4.8 years and included 24,847 male participants who were followed for a total of 119,000 person years. The average age of the men was 47.5 years.

The analysis found:

- The men receiving cholesterol reduction treatment reduced their cholesterol levels by about 10%.
- The men receiving cholesterol reduction treatment had a 7% increase in death rates compared to the men taking a placebo.
- The men receiving cholesterol reduction treatment had up to a 30% increase in deaths from accidents, suicides and homicides.

The results of this analysis of six trials show that as cholesterol levels are lowered, the death rates from accidents, suicides and homicides are increased.

Paper 53
Men with the lowest cholesterol have a three-fold increased risk of death from AIDS compared to men with the highest cholesterol
Neaton, JD et al. "Low serum cholesterol and risk of death from AIDS". *AIDS.* 1997 Jun;11(7):929-30

The study investigated the relationship of cholesterol levels, measured prior to HIV infection, to the risk of death from AIDS. The study included 332,547 men aged between 35-57, who were followed for 16 years.

This large study found that men with the lowest cholesterol levels, below 160 mg/dL (4.2 mmol/L), had a three-fold increased risk of death from AIDS compared to men who had cholesterol levels above 240 mg/dL (6.2 mmol/L).

Paper 54
Violence, anti-social behaviour and premature death associated with low cholesterol levels

Repo-Tiihonen, E et al. "Total serum cholesterol level, violent criminal offences, suicidal behavior, mortality and the appearance of conduct disorder in Finnish male criminal offenders with anti-social personality disorder". *European Archives of Psychiatry and Clinical Neuroscience.* Volume 252, Number 1, 8-11

Dr Eila Repo-Tiihonen, who headed the study, is a specialist in psychiatry and forensic psychiatry. She notes that associations between low cholesterol levels and anti-social personality disorder (ASPD), violent and suicidal behaviour have been found.

In this study, Dr Repo-Tiihonen and her colleagues looked into the associations between cholesterol levels, violent and suicidal behaviour, the age of onset of the conduct disorder and the age of death among 250 Finnish male criminal offenders with ASPD.

The researchers discovered:

- Conduct disorder had begun before the age of 10, two times more often in non-violent criminal

offenders who had lower than average cholesterol levels.
- The violent criminal offenders who had lower than average cholesterol levels were seven times more likely to die before the average age of death.
- The violent offenders who had lower than average cholesterol levels were eight times more likely to die of unnatural causes.
- The average cholesterol levels of male offenders with ASPD were lower than that of the general Finnish male population.
- Low cholesterol levels are associated with childhood onset type of conduct disorder.
- Low cholesterol levels are associated with premature and unnatural mortality among male offenders with ASPD.
- Low cholesterol levels seem to be a marker for boys with conduct disorder and anti-social male offenders.

The results of Dr Repo-Tiihonen's study show that low cholesterol is associated with violent behaviour and an early death.

Paper 55
Low cholesterol levels predict death in patients with bacteria in the blood

Richardson, JP et al. "Risk factors for the development of bacteremia in nursing home patients". *Archives of Family Medicine.* 1995 Sep;4(9):785-9

This study, which was based at the University of Maryland School of Medicine, examined the association between cholesterol levels and death rates in people with bacteremia (bacteremia is the presence of bacteria in the blood). The study included 26 patients who were admitted into a nursing home.

The study found that the only admission characteristic of patients that was associated with death caused by bacteremia was low cholesterol of 3.79 mmol/L (147 mg/dL) in patients who died, whereas patients who survived had higher cholesterol of 5.05 mmol/L (195 mg/dL).

Paper 56
Low cholesterol levels are associated with an increase in death rates especially from cancer

Wannamethee, G et al. "Low serum total cholesterol concentrations and mortality in middle-aged British men". *British Medical Journal.* 1995 Aug 12;311(7002):409-13

Dr Goya Wannamethee and colleagues at the Royal Free Hospital School of Medicine in London studied 7,735 men aged 40-59 and followed them for nearly 15 years to evaluate the relationship between low cholesterol levels and causes of mortality.

After 15 years, the study found:

- Men with low cholesterol (below 4.8 mmol/L (185 mg/dL)) had a 60% increase in total death rates compared to men with cholesterol between 4.8-5.9 mmol/L (185-228 mg/dL).
- Men with low cholesterol (below 4.8 mmol/L (185 mg/dL)) had a significant increase in cancer death

rates compared to men with cholesterol between 4.8-5.9 mmol/L (185-228 mg/dL).
- Low cholesterol levels were associated with an increased prevalence of several diseases and indicators of ill health.

The findings of this study indicate that low cholesterol levels are associated with an increase in death rates, especially from cancer.

Paper 57
Men and women with low cholesterol have higher rates of cancer deaths

Iso, H et al. "Serum total cholesterol and mortality in a Japanese population". *Journal of Clinical Epidemiology.* 1994 Sep;47(9):961-9

The study examined the relationship between cholesterol levels and cancer deaths and total deaths. Cholesterol levels were taken from 12,187 men and women aged 40-69 years living in Osaka who were then followed for nearly nine years.

The study revealed:

- Women with the lowest cholesterol had an increased risk of death and cancer deaths.
- Men with the lowest cholesterol had a significantly increased risk of death and cancer deaths.

This study found that lower cholesterol levels lead to higher levels of death and cancer death, particularly in men.

Paper 58
Colon cancer deaths increase
in men with low cholesterol

Cowan, L et al. "Cancer mortality and lipid and lipoprotein levels. The lipid research clinics program mortality follow-up study". *American Journal of Epidemiology.* Vol. 131, No. 3: 468-482

This study examined the associations of total cholesterol levels and low-density lipoprotein (LDL) cholesterol levels, with the risk of death from colon cancer in 2,753 men and 2,476 women aged 40-79 over an $8\,{}^1/_2$-year period. The senior investigator of the study was Dr Linda Cowan, who is Chair of the Department of Biostatistics and Epidemiology at the University of Oklahoma Health Science Center College of Public Health.

The study found:

- Men with the lowest LDL cholesterol (119 mg/dL or 3 mmol/L) (the so-called bad cholesterol) had a 379% increased risk of colon cancer compared with men who had higher levels of LDL.
- Men with the lowest total cholesterol (187 mg/dL or 4.8 mmol/L) had a 420% increased risk of colon cancer compared to men who had higher cholesterol levels.

Dr Cowan's findings demonstrate that low cholesterol levels lead to an increased risk of death from colon cancer.

Paper 59
Analysis of 519,643 people reveals low cholesterol
increases the risk of dying from pancreatic cancer by 27%

Ansary-Moghaddam, A et al. "The effect of modifiable risk factors on pancreatic cancer mortality in

populations of the Asia-Pacific region".
Cancer Epidemiology, Biomarkers and Prevention.
2006 Dec;15(12):2435-40

In 2011, Dr Alireza Ansary-Moghaddam was awarded the State of Kuwait prize for the control of cancer, cardiovascular diseases and diabetes in the Eastern Mediterranean region.

In this analysis of 30 studies, Dr Ansary-Moghaddam assessed the connection between cholesterol levels and death rates from pancreatic cancer. The study included 519,643 men and women with 3,558,733 person-years of follow-up.

This analysis of 519,643 people revealed that those with the lowest cholesterol, below 4.8 mmol/L (185 mg/dL), had a 27% increased risk of dying from pancreatic cancer compared to those with the highest cholesterol, over 5.8 mmol/L (224 mg/dL).

Paper 60
A rise in total cholesterol reduces the risk of death from cancer and infections in the oldest old
Weverling-Rijnsburger, AWE et al. "Total cholesterol and risk of mortality in the oldest old".
The Lancet. Volume 350, Issue 9085,
Pages 1119 - 1123, 18 October 1997

This Leiden University study assessed the influence of cholesterol levels on specific and all-cause death rates in people aged 85 years and over. The study included 724 participants (with an average age of 89 years), whose cholesterol levels were measured and death risks were calculated over 10 years of follow-up.

The participants were placed into 3 categories:

- Those with cholesterol less than 193 mg/dL (5.0 mmol/l)
- Those with cholesterol between 193 mg/dL (5.0 mmol/l) and 247 mg/dL (6.4 mmol/l)
- Those with cholesterol above 247 mg/dL (6.4 mmol/l)

The data from the study revealed:

- Each 1 mmol/L increase in total cholesterol corresponded to a 15% decrease in death rates.
- Death rates from cancer and infection were significantly lower among the participants in the highest cholesterol category than in the other categories.

The results of this Dutch study show that in people older than 85 years of age, high total cholesterol concentrations are associated with longevity and lower death rates from cancer and infection.

Paper 61
A review of 150 studies finds an association between low cholesterol and death from injury

Cummings, P et al. "The association between cholesterol and death from injury". *Annals of Internal Medicine.* 1994 May 15;120(10):848-55

The author of this paper, Peter Cummings, is Emeritus Professor of Epidemiology at the University of

Washington. In this paper, he reviewed over 150 studies concerning the association between low cholesterol and death from injury.

The review found:

- Men whose cholesterol levels were lower than 4.14 mmol/L (160 mg/dL) had a 40% higher risk of death from injury compared with men whose cholesterol levels were 4.14 to 5.15 mmol/L (160 to 200 mg/dL).
- In trials where the objective was cholesterol reduction, those who were treated with cholesterol-reducing regimes had a 42% higher risk of death from injury compared with those who had no treatment.

Dr Cummings' review of 150 studies found an association between low cholesterol and death from injury.

Paper 62
Low cholesterol leads to increased rates of deaths from cancer and injuries, and deaths from diseases of the respiratory and digestive systems

Hulley, SB et al. "Health policy on blood cholesterol. Time to change direction". *Circulation*. 1992;86;1026-1029

Dr Stephen Hulley is lead author of the widely used textbook *Designing Clinical Research*. With 100,000 copies sold, it is now in its third edition (2007).

Dr Hulley reviewed the study "Report of the Conference on Low Blood Cholesterol: Mortality Associations" that included 68,406 deaths.

He found:

- Men with cholesterol below 160 mg/dl had a 20% higher rate of cancer deaths compared with those with higher cholesterol levels.
- Men with cholesterol below 160 mg/dl had a 40% higher rate of non-cardiovascular, non-cancer deaths compared with those with higher cholesterol levels.
- Men with cholesterol below 160 mg/dl had a 35% increased rate of injury deaths compared with those with higher cholesterol levels.
- Men with cholesterol below 160 mg/dl had a 15% increased rate of respiratory system deaths compared with those with higher cholesterol levels.
- Men with cholesterol below 160 mg/dl had a 50% increased rate of digestive system deaths compared with those with higher cholesterol levels.
- Among women, the patterns of the association between low blood cholesterol and increased rates of various causes of non-cardiovascular deaths were similar to those in men, except that the excess in cancer mortality was smaller (about 5%).

The results of the study show that low cholesterol leads to increased rates of deaths from cancer and injuries, and deaths from diseases of the respiratory and digestive systems.

Paper 63

Low levels of cholesterol are predictive of higher rates of death in patients with end-stage renal disease

Bowden, RG et al. "Reverse epidemiology of lipid-death associations in a cohort of end-stage renal disease patients". *Nephron Clinical Practice.* 2011;119(3):c214-9

End-stage kidney disease is the complete or almost complete failure of the kidneys to work. "Reverse epidemiology" means that obesity and high cholesterol may be protective and associated with greater survival in disease.

The purpose of this study was to determine if there are reverse epidemiological associations between cholesterol and mortality in end-stage renal disease patients. 438 patients with end-stage renal disease were tracked for 36 months.

The study found that low levels of cholesterol and low levels of low-density lipoprotein (LDL) cholesterol were predictive of higher rates of death in patients with end-stage renal disease.

Paper 64
Kidney failure patients live longer
if they have high cholesterol
Iseki, K et al. "Hypocholesterolemia is a significant predictor of death in a cohort of chronic hemodialysis patients". *Kidney International.* 2002 May;61(5):1887-93

This Japanese study investigated the impact of cholesterol levels in kidney failure patients receiving hemodialysis treatment. The study included 1,167 hemodialysis patients, who were followed for ten years.

The study revealed that 12.7% more patients survived who had the highest cholesterol, above 220 mg/dL (5.7 mmol/L), compared to the patients with the lowest cholesterol, below 140 mg/dL (3.6 mmol/L).

The findings of the study show that low cholesterol is a significant predictor of death in kidney failure patients receiving hemodialysis treatment.

Paper 65
Higher risk of liver diseases with low cholesterol

Chen, Z et al. "Prolonged infection with hepatitis B virus and association between low blood cholesterol concentration and liver cancer". *British Medical Journal.* 1993 Apr 3;306(6882):890-4

Hepatitis can be caused by viruses that primarily attack the liver cells, such as hepatitis B. About one-fifth of patients with chronic hepatitis B are at risk of developing cirrhosis or cancer of the liver.

In this study, Professor Zhengming Chen and his team from the University of Oxford examined 1,556 apparently healthy men aged 35-64 years to determine whether prolonged infection with hepatitis B virus is associated with a lower blood cholesterol concentration. 238 (15%) of the men were positive for hepatitis B surface antigen, indicating that they were chronic carriers.

The researchers discovered:

- Cholesterol levels were 4.2% lower among carriers (that is, positive for hepatitis B surface antigen) than among non-carriers.
- Chronic hepatitis B virus infection, which usually starts in early childhood in China, seems to lead not only to a greatly increased risk of death from liver

disease, but also to a somewhat lower cholesterol levels in adulthood.

The results of the study indicate that lower cholesterol levels lead to risk of death from liver cancer and from other chronic liver diseases.

Paper 66
The connection between chronic obstructive pulmonary disease and low cholesterol

Sin, DD et al. "Why are patients with chronic obstructive pulmonary disease at increased risk of cardiovascular diseases? The potential role of systemic inflammation in chronic obstructive pulmonary disease". *Circulation.* 2003 Mar 25;107(11):1514-9

This study analysed the association of cholesterol levels with the severity of airflow obstruction in people with chronic obstructive pulmonary (lung) disease. The study included 6,629 people aged 50 or over.

Dr Don Sin was the head investigator of the study. His main research focus is to improve the care and diagnosis of patients with chronic obstructive pulmonary disease. Dr Sin notes that airflow obstructions elevate the risk of ischemic heart diseases, strokes, and sudden cardiac deaths two- to three-fold.

The study found:

- People with no airflow obstruction had the highest cholesterol.
- People with the most severe airflow obstruction had the lowest cholesterol.

Dr Sin's research indicates that low cholesterol is associated with a higher severity of chronic obstructive pulmonary disease, which may lead to a two- to three-fold increased risk of ischemic heart disease, stroke, and sudden cardiac death.

Paper 67
Low cholesterol levels are associated with higher death rates from respiratory diseases

Iribarren, C et al. "Serum total cholesterol and risk of hospitalisation, and death from respiratory disease". *International Journal of Epidemiology*. 1997 Dec;26(6):1191-202

This study examined the association of cholesterol levels with respiratory diseases. The study included 48,188 men and 55,276 women with an age range of 25-89, who were followed for 15 years, with a total of 976,866 person years of observation.

The study found that for patients requiring hospitalisation:

- Those with the lowest cholesterol levels, below 4.14 mmol/L (160 mg/dL), had a 41% increased risk of been hospitalised with *pneumonia and influenza* compared with those with the highest cholesterol levels, above 6.2 mmol/L (240 mg/dL).
- Those with the lowest cholesterol levels, below 4.14 mmol/L (160 mg/dL), had a 17% increased risk of been hospitalised with chronic obstructive pulmonary disease (*bronchitis and emphysema*) compared with those with the highest cholesterol levels, above 6.2 mmol/L (240 mg/dL).

- Those with the lowest cholesterol levels, below 4.14 mmol/L (160 mg/dL), had a 13% increased risk of been hospitalised with *asthma* compared with those with the highest cholesterol levels, above 6.2 mmol/L (240 mg/dL).
- Those with the lowest cholesterol levels, below 4.14 mmol/L (160 mg/dL), had a 35% increased risk of been hospitalised with other respiratory diseases (*rhinitis, sinusitis, tonsillitis, laryngitis, asbestosis, pneumuconiosis, empyema, mediastinitis, pleurisy, pulmonary congestion, pulmonary fibrosis, rhumatic pneumonia and lung disease*) compared with those with the highest cholesterol levels, above 6.2 mmol/L (240 mg/dL).

With regard to death from respiratory diseases, the study found:

- Men with the lowest cholesterol levels, below 4.14 mmol/L (160 mg/dL), had a 87% increased risk of death from *pneumonia and influenza* compared with men with the highest cholesterol levels, above 6.2 mmol/L (240 mg/dL).
- Women with the lowest cholesterol levels, below 4.14 mmol/L (160 mg/dL), had a 41% increased risk of death from *pneumonia and influenza* compared with women with the highest cholesterol levels, above 6.2 mmol/L (240 mg/dL).
- Men with the lowest cholesterol levels, below 4.14 mmol/L (160 mg/dL), had a 35% increased risk of death from *bronchitis, emphysema and asthma* compared with men with the highest cholesterol levels, above 6.2 mmol/L (240 mg/dL).

- Women with the lowest cholesterol levels, below 4.14 mmol/L (160 mg/dL), had a 79% increased risk of death from *bronchitis, emphysema and asthma* compared with women with the highest cholesterol levels, above 6.2 mmol/L (240 mg/dL).
- Men with the lowest cholesterol levels, below 4.14 mmol/L (160 mg/dL), had a 96% increased risk of death from other respiratory diseases (*rhinitis, sinusitis, tonsilitis, laringitis, asbestosis, pneumuconiosis, empyema, mediastinitis, pleurisy, pulmonary congestion, pulmonary fibrosis, rhumatic pneumonia and lung disease*) compared with men with the highest cholesterol levels, above 6.2 mmol/L (240 mg/dL).
- Women with the lowest cholesterol levels, below 4.14 mmol/L (160 mg/dL), had a 126% increased risk of death from other respiratory diseases (*rhinitis, sinusitis, tonsilitis, laringitis, asbestosis, pneumuconiosis, empyema, mediastinitis, pleurisy, pulmonary congestion, pulmonary fibrosis, rhumatic pneumonia and lung disease*) compared with women with the highest cholesterol levels, above 6.2 mmol/L (240 mg/dL).

The results of this 15-year study show that low cholesterol levels are associated with more hospitalisations and higher death rates from respiratory diseases.

Paper 68
Correlation between low cholesterol levels and rheumatoid arthritis

Hurt-Camejo, E et al. "Elevated levels of small, low-density lipoprotein with high affinity for arterial matrix components in patients with rheumatoid arthritis: possible contribution of phospholipase

A2 to this atherogenic profile". *Arthritis and Rheumatism.* 2001 Dec;44(12):2761-7

The principle investigator of this study was Eva Hurt-Camejo, who is a Professor in Vascular Biology in the Faculty of Medicine at the University of Gothenburg, Sweden. She notes that increased death rates and premature death due to cardiovascular disease are more common in patients with rheumatoid arthritis compared with the general population.

The study investigated the association of cholesterol levels with rheumatoid arthritis. The study included 31 rheumatoid arthritis patients and 28 control subjects. Average age of the study participants was 53 years.

Professor Hurt-Camejo found that rheumatoid arthritis patients (with their higher death rates) had 6.5% lower cholesterol levels than the control subjects.

Paper 69
Meningococcal sepsis is associated with low cholesterol levels

Vermont, CL et al. "Serum lipids and disease severity in children with severe meningococcal sepsis". *Critical Care Medicine.* 2005 Jul;33(7):1610-5

Meningococcal sepsis is where bacteria have invaded the bloodstream. This results in fever, irritability, headaches and a stiff neck. Once the bacteria are in the blood, they begin to attack organs and cause internal bleeding, with potentially fatal results within a matter of hours.

The aim of this Rotterdam-based study was to evaluate the role of cholesterol levels in children with severe meningococcal sepsis. The study included 57 patients admitted to the paediatric intensive care unit with meningococcal sepsis or septic shock.

After analysing the patients, the researchers found:

- Cholesterol levels on admission to the paediatric intensive care unit were very low in all patients.
- Cholesterol levels were significantly lower in non-survivors than in survivors.
- The lower the cholesterol levels, the more severe the illness.

The results of this Dutch study show that low cholesterol levels are associated with meningococcal sepsis and the lower the cholesterol is, the more severe the disease is.

Paper 70
Both low cholesterol levels and declining cholesterol levels are associated with increased risk of death from suicide in men

Zureik, M et al. "Serum cholesterol concentration and death from suicide in men: Paris prospective study I". *British Medical Journal.* 1996 Sep 14;313(7058):649-51

The aim of the study, carried out by Dr Mahmoud Zureik and his team in Paris, was to investigate whether low cholesterol levels or changing cholesterol levels are associated with the risk of suicide in men.

The study assessed 6,393 working men, aged 43-52, who had at least three measurements of their cholesterol levels taken over a 17-year period.

Dr Zureik found:

- Men with low cholesterol (below 4.78 mmol/L (185 mg/dL)) had a 216% increase in suicide compared with men whose cholesterol was between 4.78-6.21 mmol/L (185-240mg/dL).
- Men whose cholesterol levels decreased by more than 0.13 mmol/L (5 mg/dL) a year had a 117% increase in suicide compared with men whose cholesterol remained stable.

This study reveals that both low cholesterol levels and declining cholesterol levels are associated with increased risk of death from suicide in men.

Paper 71
Those with the lowest cholesterol levels have more than six times the risk of committing suicide than those with the highest cholesterol levels

Ellison, L et al. "Low Serum Cholesterol Concentration and Risk of Suicide". *Epidemiology.* March 2001 - Volume 12 - Issue 2 - pp 168-172

This Canadian study examined the relationship between low cholesterol levels and mortality from suicide. The study included 11,554 people aged between 11 and 84 who were followed for an average of 12 years.

The study results showed that those with the lowest cholesterol levels, under 4.27 mmol/L (165 mg/dL), had more than six times the risk of committing suicide, as did subjects with the highest cholesterol levels, over 5.77 mmol/L (223 mg/dL).

Paper 72
The lower the cholesterol level, the higher the risk of suicide
Lee, HJ et al. "Serum lipid levels and suicide attempts". *Acta Psychiatrica Scandinavica.* 2003 Sep; 108(3): 215-21

The objective of the study was to determine whether a correlation exists between lower cholesterol levels and increased suicide risk. Cholesterol levels were measured in 60 patients who had recently experienced failed attempts at suicide and equal numbers of non-suicidal patients.

The study revealed:

- Total cholesterol levels were significantly lower in suicide attempt patients compared to non-suicidal patients.
- Low-density lipoprotein (LDL) cholesterol levels were significantly lower in suicide attempt patients compared to non-suicidal patients.
- The lower the cholesterol level. The higher the risk of suicide.

This study shows the lower the cholesterol and LDL levels, the higher the suicide risk.

Paper 73
Rising cholesterol levels predict survival, whilst falling cholesterol levels predicts death in people who have suffered severe injuries
Dunham, CM et al. "Following severe injury, hypocholesterolemia improves with convalescence

but persists with organ failure or onset of infection".
Critical Care. 2003 Dec;7(6):R145-53

The lead author of the study, Dr Michael Dunham, is a surgeon at the Elizabeth Health Center in Ohio. The objective of the study was to determine the association of cholesterol levels and severe traumatic injury. The study included 28 severely injured trauma patients.

The study found:

- The cholesterol levels in those who survived their severe injuries increased by 28%.
- The cholesterol levels of those who died of their severe injuries decreased by 33%.

Dr Dunham's research reveals that a rising cholesterol level predicts survival, and a falling cholesterol level predicts death in people who have suffered severe injuries.

Paper 74
Men with low cholesterol are susceptible to cancer and tuberculosis
Kozarevic, DJ et al. "Serum cholesterol and mortality. The Yugoslavia cardiovascular disease study. *American Journal of Epidemiology.* Vol. 114, No. 1: 21-28 1981"

This study, published in the *American Journal of Epidemiology,* investigated the relationship of cholesterol levels to the 7-year incidence of death from all causes and from specific causes in 11,121 Yugoslav males aged 35-62.

The study found:

- The men with the lowest cholesterol, below 150 mg/dl (3.9 mmol/l), had a 21% higher death rate than the men with the highest cholesterol, above 270 mg/dL (7.0 mmol/l).
- Men with low cholesterol were susceptible to death from cancer and particularly to death from respiratory diseases such as tuberculosis.

The results of the study demonstrate that if you have low cholesterol levels, it is significantly likely you will die earlier, especially from cancer and respiratory diseases.

Chapter four has exposed the dangers of low cholesterol and how it may predispose individuals to death in a myriad of illnesses and conditions, ranging from asthma to kidney disease and from pancreatic cancer to suicide.

But what is the influence of cholesterol on our health when we are alive? Are we healthier throughout life with low cholesterol or high cholesterol?

Chapter five explores the influence of cholesterol levels on 77 diseases and conditions.

CHAPTER 5

Low cholesterol leads to an increased prevalence of many diseases

The previous four chapters contained 74 scientific papers that demonstrated how low cholesterol levels may lead to an early death.

But during our time on Earth, we may ask: does low cholesterol help us to be strong and healthy? Will low cholesterol make us more intelligent? What are the effects of low cholesterol on our mental health? Do we suffer less minor illness with low cholesterol? Does low cholesterol make us happy and joyous? Will low cholesterol enable us to enter old age without suffering from a debilitating neurological disorder? Will low cholesterol protect us from infection? And will low cholesterol make us vibrant and full of life?

In this chapter, we explore the effects of cholesterol as we journey through life.

How do cholesterol levels affect our health from the day we are born?

What about its influence as we enter our teens and then move into young adulthood?

When middle age beckons, and we start to have health concerns, what does cholesterol do for us?

As we approach old age, should we aim for low or high cholesterol to help us enjoy many flourishing robust retirement years?

The following chapter answers this plethora of questions.

Paper 75
After childbirth, women with lower cholesterol levels have major feelings of fatigue and a depressed mood

Nasta, MT et al. "Cholesterol and mood states at 3 days after delivery". *Journal of Psychosomatic Research*. 2002 Feb;52(2):61-3

The aim of the study, based at the University of Padua, was to examine the association between cholesterol and mood states in women immediately following childbirth. The study included 72 women who had their mood analysed and cholesterol levels measured three days after giving birth.

This Italian study found that following childbirth women with lower cholesterol levels had major feelings of fatigue and a depressed mood.

Paper 76
The link between low cholesterol and autism

Aneja, A et al. "Autism: the role of cholesterol in treatment". *International Review of Psychiatry*. 2008 Apr;20(2):165-70

Doctor Alka Aneja from the Johns Hopkins University School of Medicine presided over this review concerning autism and cholesterol. Dr Aneja notes that cholesterol is essential for normal embryonic and fetal development and a deficit of cholesterol may contribute to autism spectrum disorders, as observed in Smith-Lemli-Opitz syndrome. Smith-Lemli-Opitz syndrome is a disorder where the body fails to produce enough cholesterol.

The review found that Smith-Lemli-Opitz syndrome is treatable by increasing dietary cholesterol.

Individuals with Smith-Lemli-Opitz syndrome who have such cholesterol treatment display fewer autistic behaviours, infections, and symptoms of irritability and hyperactivity, with improvements in physical growth, sleep and social interactions.

Other behaviours shown to improve by increasing cholesterol include aggressive behaviours, self-injury, temper outbursts and trichotillomania (pulling out one's own hair).

Dr Aneja's review found that low cholesterol is associated with autism and an increase in dietary cholesterol helps to alleviate autistic behaviour.

Paper 77
Low cholesterol values may be a risk factor for aggression in children

Zhang, J et al. "Association of Serum Cholesterol and History of School Suspension among School-age Children and Adolescents in the United States".
American Journal of Epidemiology. (2005) 161 (7): 691-699

This study, led by Jian Zhang, an Associate Professor of the Department of Biology at the University of North Carolina, set out to determine the association of cholesterol levels with pupils who had been suspended or expelled from school. The study included 4,852 children aged 6–16 years.

The study showed that children between 6 and 16 years of age whose cholesterol concentration was below 3.77 mmol/L (145 mg/dL) were 73% more likely to have been suspended or expelled from schools than children with higher cholesterol levels.

Professor Zhang concluded that low total cholesterol is associated with school suspension or expulsion and that low cholesterol may be a risk factor for aggression.

Paper 78
Low cholesterol is significantly associated with schizophrenia

Atmaca, M et al. "Serum leptin and cholesterol levels in schizophrenic patients with and without suicide attempts". *Acta Psychiatrica Scandinavica.* 2003 Sep;108(3):208-14

This study examined the relationship of cholesterol levels in schizophrenic patients with and without suicide attempts. The study included 16 medication-free schizophrenic patients with and without suicide attempts and 16 healthy controls.

The study found:

- The schizophrenic patients had lower cholesterol levels compared with the controls.

- The schizophrenic patients who had attempted suicide had significantly lower cholesterol levels compared with the schizophrenic patients who had not.
- Cholesterol levels were lower in violent suicide attempters when compared with non-violent suicide attempters.

The results of the study show that low cholesterol is significantly associated with schizophrenia, especially in suicide attempters and even more so in violent suicide attempters.

Paper 79
People with low cholesterol are more prone to coughs, colds, runny noses and sore throats
Hyman, DJ et al. "Effect of minor illness on serum cholesterol level". *American Journal of Preventive Medicine.* 1992 Mar-Apr;8(2):100-3

This study, based at the prestigious Stanford University, investigated the association of cholesterol levels with a minor illness. The study included 6,880 people who had their cholesterol levels measured. Additionally, a further 162 people were followed for six years. Minor illnesses include coughs, colds, runny noses and sore throats.

In the main study:

- Those who had a minor illness on the day their cholesterol was measured had 2.7% lower cholesterol levels compared to healthy subjects.
- Those who had a minor illness on the day their high-density lipoprotein (HDL) cholesterol was

measured had 2.1% lower HDL cholesterol levels compared to healthy subjects.

In the six-year study, those with a minor illness had 2.2% lower cholesterol levels compared to healthy subjects.

The results of this Stanford study indicate that people with low cholesterol are more prone to minor illnesses such as coughs, colds, runny noses and sore throats.

Paper 80
Crohn's disease and ulcerative colitis link with low cholesterol

Ripollés, Piquer B et al. "Altered lipid, apolipoprotein, and lipoprotein profiles in inflammatory bowel disease: consequences on the cholesterol efflux capacity of serum using Fu5AH cell system". *Metabolism.* 2006 Jul;55(7):980-8

This French study examined the link between cholesterol levels with inflammatory bowel diseases such as Crohn's disease and ulcerative colitis. The study included 21 patients with inflammatory bowel diseases and 28 healthy subjects.

The study found that the patients with inflammatory bowel diseases had significantly lower levels of cholesterol and high-density lipoprotein (HDL) cholesterol than the healthy subjects.

Paper 81
Low cholesterol levels are associated with higher rates of many infectious diseases

Iribarren, C et al. "Cohort study of serum total cholesterol and in-hospital incidence

of infectious diseases". *Epidemiology and Infection.*
1998 Oct;121(2):335-47

This study assessed the association between cholesterol levels and the risk of infections (other than respiratory and HIV) requiring hospitalisation. The study included 55,300 men and 65,271 women, who were followed for 15 years.

The infectious diseases analysed in this study were classified as follows:

- Intestinal infections: salmonella, rotavirus (a viral infection that can cause gastroenteritis)
- Viral hepatitis
- Acute appendicitis
- All digestive and liver infections: diverticulosis, abscess of the anal and rectal region, abscess of the intestine, liver disease, gallbladder inflammation, cholangitis (bile duct infection)
- Endocarditis (inflammation of the inner layer of the heart)
- Kidney infections
- Urinary tract infections
- All genito-urinary infections: cystitis, prostatitis, orchitis and epididymitis (inflammation of the testicles).
- Venereal diseases: syphilis, gonorrhoea, chlamydia, trichomoniasis
- Musculo-skeletal infections: arthropathy (disease of the joints), infective myositis (skeletal muscle infection), osteomyelitis (infection of the bone or bone marrow), periostitis (inflammation of the

periosteum, a layer of connective tissue that surrounds bone)
- Skin and subcutaneous tissue: herpes, eczema, ringworm, thrush, carbuncle, boils, cellulitis (common skin infection caused by bacteria), lymphadenitis (swollen or enlarged lymph nodes), impetigo (bacterial skin infection), pilonidal cyst (cyst or abscess under the skin of the buttocks), pyoderma (skin infection that exudes pus)
- Septicaemia, bacteraemia
- Gangrene
- Central and peripheral nervous system: meningitis, encephalitis (inflammation of the brain), myelitis (inflammation of the spinal cord), abscess on the brain, abscess on the spinal cord
- Endotoxic shock (septic shock)
- Gynaecological: salpingitis (infection and inflammation in the fallopian tubes), oophoritis (inflammation of the ovaries), pelvic inflammatory disease (bacterial infection of the female upper genital tract, including the womb, fallopian tubes and ovaries), cervicitis (inflammation of the uterine cervix), vaginitis (inflammation of the vagina), bartholin cyst or abscess, (infection of the bartholin's glands which lie next to the entrance to the vagina)

The study found for men:

- Men with lower cholesterol levels below 4.14 mmol/L (160 mg/dL) had a 29% increased risk of being hospitalised with *any infection* compared to men with higher cholesterol levels over 6.2 mmol/L (240 mg/dL).

- Men with lower cholesterol levels below 4.14 mmol/L (160 mg/dL) had a 9% increased risk of being hospitalised with *intestinal infections* compared to men with higher cholesterol levels between 4.14-5.15 mmol/L (160-200 mmol/L).
- Men with lower cholesterol levels below 4.14 mmol/L (160 mg/dL) had a 63% increased risk of being hospitalised with *viral hepatitis* compared to men with higher cholesterol levels over 6.2 mmol/L (240 mg/dL).
- Men with lower cholesterol levels below 4.14 mmol/L (160 mg/dL) had a 1% increased risk of being hospitalised with *acute appendicitis* compared to men with higher cholesterol levels over 6.2 mmol/L (240 mg/dL).
- Men with lower cholesterol levels below 4.14 mmol/L (160 mg/dL) had a 15% increased risk of being hospitalised with *all digestive and liver infections* compared to men with higher cholesterol levels over 6.2 mmol/L (240 mg/dL).
- Men with lower cholesterol levels between 4.14-5.15 mmol/L (160-200 mmol/L) had a 22% increased risk of being hospitalised with *endocarditis* compared to men with higher cholesterol levels over 6.2 mmol/L (240 mg/dL).
- Men with lower cholesterol levels below 4.14 mmol/L (160 mg/dL) had a 10% increased risk of being hospitalised with *kidney infections* compared to men with higher cholesterol levels between 5.16-6.19 mmol/L (200-240 mg/dL).
- Men with lower cholesterol levels below 4.14 mmol/L (160 mg/dL) had a 27% increased risk of being hospitalised with *urinary tract infections*

compared to men with higher cholesterol levels over 6.2 mmol/L (240 mg/dL).
- Men with lower cholesterol levels below 4.14 mmol/L (160 mg/dL) had a 18% increased risk of being hospitalised with all *genito-urinary infections* compared to men with higher cholesterol levels over 6.2 mmol/L (240 mg/dL).
- Men with lower cholesterol levels below 4.14 mmol/L (160 mg/dL) had a 140% increased risk of being hospitalised with *venereal diseases* compared to men with higher cholesterol levels over 6.2 mmol/L (240 mg/dL).
- Men with lower cholesterol levels below 4.14 mmol/L (160 mg/dL) had a 28% increased risk of being hospitalised with *musculo-skeletal infections* compared to men with higher cholesterol levels over 6.2 mmol/L (240 mg/dL).
- Men with lower cholesterol levels below 4.14 mmol/L (160 mg/dL) had a 22% increased risk of being hospitalised with *skin and subcutaneous tissue infections* compared to men with higher cholesterol levels over 6.2 mmol/L (240 mg/dL).
- Men with lower cholesterol levels below 4.14 mmol/L (160 mg/dL) had a 17% increased risk of being hospitalised with *septicaemia and bacteraemia* compared to men with higher cholesterol levels over 6.2 mmol/L (240 mg/dL).
- Men with lower cholesterol levels below 4.14 mmol/L (160 mg/dL) had a 20% increased risk of being hospitalised with *gangrene* compared to men with higher cholesterol levels between 5.16-6.19 mmol/L (200-240 mg/dL).
- Men with lower cholesterol levels below 4.14 mmol/L (160 mg/dL) had a 67% increased risk of

being hospitalised with *central and peripheral nervous system infections* compared to men with higher cholesterol levels over 6.2 mmol/L (240 mg/dL).
- Men with lower cholesterol levels below 4.14 mmol/L (160 mg/dL) had a 128% increased risk of being hospitalised with *endotoxic shock* compared to men with higher cholesterol levels over 6.2 mmol/L (240 mg/dL).

For women, the study found:

- Women with lower cholesterol levels below 4.14 mmol/L (160 mg/dL) had a 30% increased risk of being hospitalised with *any infection* compared to women with higher cholesterol levels over 6.2 mmol/L (240 mg/dL).
- Women with lower cholesterol levels below 4.14 mmol/L (160 mg/dL) had a 67% increased risk of being hospitalised with *intestinal infections* compared to women with higher cholesterol levels over 6.2 mmol/L (240 mg/dL).
- Women with lower cholesterol levels below 4.14 mmol/L (160 mg/dL) had a 51% increased risk of being hospitalised with *viral hepatitis* compared to women with higher cholesterol levels over 6.2 mmol/L (240 mg/dL).
- Women with lower cholesterol levels below 4.14 mmol/L (160 mg/dL) had a 22% increased risk of being hospitalised with *acute appendicitis* compared to women with higher cholesterol levels over 6.2 mmol/L (240 mg/dL).
- Women with lower cholesterol levels below 4.14 mmol/L (160 mg/dL) had a 19% increased risk of being hospitalised with all *digestive and liver infections*

compared to women with higher cholesterol levels over 6.2 mmol/L (240 mg/dL).
- Women with lower cholesterol levels below 4.14 mmol/L (160 mg/dL) had a 137% increased risk of being hospitalised with *endocarditis* compared to women with higher cholesterol levels over 6.2 mmol/L (240 mg/dL).
- Women with lower cholesterol levels below 4.14 mmol/L (160 mg/dL) had a 9% increased risk of being hospitalised with *kidney infections* compared to women with higher cholesterol levels over 6.2 mmol/L (240 mg/dL).
- Women with lower cholesterol levels between 4.14-5.15 mmol/L (160-200 mmol/L) had a 28% increased risk of being hospitalised with *urinary tract* infections compared to women with higher cholesterol levels over 6.2 mmol/L (240 mg/dL).
- Women with lower cholesterol levels below 4.14 mmol/L (160 mg/dL) had a 33% increased risk of being hospitalised with all *genito-urinary infections* compared to women with higher cholesterol levels over 6.2 mmol/L (240 mg/dL).
- Women with lower cholesterol levels below 4.14 mmol/L (160 mg/dL) had a 90% increased risk of being hospitalised with *venereal diseases* compared to women with higher cholesterol levels over 6.2 mmol/L (240 mg/dL).
- Women with lower cholesterol levels between 5.16-6.19 mmol/L (200-239 mg/dL) had a 4% increased risk of being hospitalised with *musculo-skeletal infections* compared to women with higher cholesterol levels over 6.2 mmol/L (240 mg/dL).

- Women with lower cholesterol levels below 4.14 mmol/L (160 mg/dL) had a 15% increased risk of being hospitalised with *skin and subcutaneous tissue* infections compared to women with higher cholesterol levels over 6.2 mmol/L (240 mg/dL).
- Women with lower cholesterol levels below 4.14 mmol/L (160 mg/dL) had a 39% increased risk of being hospitalised with *septicaemia and bacteraemia* compared to women with higher cholesterol levels over 6.2 mmol/L (240 mg/dL).
- Women with lower cholesterol levels below 4.14 mmol/L (160 mg/dL) had a 93% increased risk of being hospitalised with *gangrene* compared to women with higher cholesterol levels between 4.14-5.15 mmol/L (160-200 mmol/L).
- Women with lower cholesterol levels between 5.16-6.19 mmol/L (200-239 mg/dL) had a 9% increased risk of being hospitalised with *central and peripheral nervous system infections* compared to women with higher cholesterol levels over 6.2 mmol/L (240 mg/dL).
- Women with lower cholesterol levels between 5.16-6.19 mmol/L (200-239 mg/dL) had a 44% increased risk of being hospitalised with *endotoxic shock* compared to women with higher cholesterol levels over 6.2 mmol/L (240 mg/dL).
- Women with lower cholesterol levels below 4.14 mmol/L (160 mg/dL) had a 2% increased risk of being hospitalised with *gynaecological infections* compared to women with higher cholesterol levels between 5.16-6.19 mmol/L (200-240 mg/dL).

The results of this study show that low cholesterol levels are associated with higher rates of many infectious diseases.

Paper 82
Hepatitis C associated with low cholesterol levels

Siagris, D et al. "Serum lipid pattern in chronic hepatitis C: histological and virological correlations". *Journal of Viral Hepatitis.* 2006 Jan;13(1):56-61

Dr Dimitrios Siagris from Patras University Hospital led this study, which investigated the relationship of cholesterol levels with the hepatitis C virus. The study included 155 patients with hepatitis C and 138 normal subjects.

This Greek study found that the patients with hepatitis C had significantly lower levels of cholesterol, high-density lipoprotein (HDL) cholesterol and low-density lipoprotein (LDL) cholesterol than the normal subjects.

Paper 83
High cholesterol DECREASES the risk of intermittent claudication by 10%

Passons, VM et al. "The Bambuí Health and Aging Study (BHAS). Prevalence of Intermittent Claudication in the Aged Population of the Community of Bambuí and its Associated Factors". *Arquivos Brasileiros de Cardiologia.* vol.77 no.5 São Paulo Nov. 2001

This study compared the association of cholesterol levels with the prevalence of intermittent claudication (leg cramps). The study included 1,485 subjects over the age of 60.

The study found that those with the lowest cholesterol (under 200 mg/dL or 5.2 mmol/L) had an increased 10%

risk of intermittent claudication compared with those with the highest cholesterol (over 240 mg/dL or 6.2 mmol/L).

Paper 84
Men with gout have lower cholesterol than healthy men

Jacobelli, S et al. "Cholesterol distribution among lipoprotein fractions in patients with gout and normal controls". *Journal of Rheumatology*. 1986 Aug;13(4):774-7

This study investigated the association of cholesterol levels with gout. The study included 29 gouty men and 34 healthy controls.

The results of the study showed that gouty patients had lower levels of total cholesterol, lower levels of low-density lipopritein (LDL) cholesterol and lower levels of high-density lipoprotein (HDL) cholesterol than the healthy men.

Paper 85
Poor memory is associated with low cholesterol

Singh-Manoux, A et al. "Low HDL cholesterol is a risk factor for deficit and decline in memory in midlife: the Whitehall II study". *Arteriosclerosis, Thrombosis, and Vascular Biology*. 2008 Aug;28(8):1556-62

Dr Archana Singh-Manoux, who was head of the study, is an expert in the social, behavioural and biological determinants of cognitive aging. In this study, Dr Singh-Manoux analysed the association of cholesterol levels with memory. 3,673 male and female participants, with an average age of 61, were subjected to a short-term verbal memory test.

Dr Singh-Manoux found:

- Poor memory was found in 53% more subjects with the lowest levels of high-density lipoprotein (HDL) cholesterol, below 40 mg/dL (1.0 mmol/L), compared to those with the highest levels of high-density lipoprotein (HDL) cholesterol, above 60 mg/dL (1.5 mmol/L).
- Poor memory was found in 10% more subjects with the lowest levels of cholesterol, below 200 mg/dL (5.1 mmol/L), compared to those with the highest levels of cholesterol 240 mg/dL (6.2 mmol/L).

The findings of this French study indicate that low cholesterol levels are associated with poor memory.

Paper 86
High cholesterol levels are associated with higher intelligence

Elias, PK et al. "Serum Cholesterol and Cognitive Performance in the Framingham Heart Study". *Psychosomatic Medicine.* 67:24-30 2005

Cognitive performance is a measurement of learning, memory, attention/concentration, abstract reasoning, concept formation, and organisational abilities.

The objective of this study, authored by Dr Penelope Elias from Boston University, was to examine the relationship between cholesterol levels and cognitive performance. Cognitive tests were administered to 789 men and 1,105 women who were free of dementia and stroke and were then followed over a 16- to 18-year period.

Dr Elias found those with the highest cholesterol levels scored 80% higher in the cognitive tests compared to those with the lowest cholesterol levels.

The data from this study shows that higher cholesterol levels are associated with higher intelligence.

Paper 87
High levels of cholesterol significantly decrease the risk of Parkinson's

Lonneke, M et al. "Serum Cholesterol Levels and the Risk of Parkinson's Disease". *American Journal of Epidemiology.* (15 November 2006) 164 (10): 998-1002

This study, published in the *American Journal of Epidemiology*, set out to determine the association between levels of cholesterol and the risk of Parkinson's disease among 6,465 subjects aged 55 or more years, with 9.4 years of follow-up.

After analysing the data, the researchers found:

- Those with the highest cholesterol levels, above 7.4 mmol/L (286 mg/dL), had a 45% decreased risk of Parkinson's disease compared to those with the lowest cholesterol, below 5.9 mmol/L (228 mmol/L).
- Men with the highest cholesterol levels had a 14% decreased risk of Parkinson's disease compared to men with the lowest cholesterol.

- Women with the highest cholesterol levels had a 84% decreased risk of Parkinson's disease compared to women with the lowest cholesterol.

The findings of the study demonstrate that high cholesterol levels significantly decrease the risk of Parkinson's disease.

Paper 88
High cholesterol levels are associated with a decreased risk of Alzheimer's disease
Reitz, C et al. "Association of higher levels of high-density lipoprotein cholesterol in elderly individuals and lower risk of late-onset Alzheimer's disease". *Archives of Neurology.* 2010 Dec;67(12):1491-7

The study was headed by Dr Christiane Reitz, whose research focuses on the identification of genetic and non-genetic factors that contribute to Alzheimer's disease. This study examined the association of cholesterol levels with Alzheimer's disease. The study included 1,130 men and women (who were free of cognitive impairment at the start of the study) aged 65 or over, with 4,469 person-years of follow-up.

The study revealed:

- Those with the highest total cholesterol levels had a 60% decreased risk of developing Alzheimer's compared to those with the lowest total cholesterol levels.
- Those with the highest high-density lipoprotein (HDL) cholesterol levels had a 60% decreased

risk of developing Alzheimer's compared to those with the lowest high density lipoprotein (HDL) cholesterol levels.
- Those with the highest low-density lipoprotein (LDL) cholesterol levels had a 40% decreased risk of developing Alzheimer's compared to those with the lowest low-density lipoprotein (LDL) cholesterol levels.

The results of Dr Reitz's study show that high levels of total cholesterol, HDL cholesterol LDL cholesterol are associated with a decreased risk of Alzheimer's disease.

Paper 89
Low cholesterol levels are associated with higher rates of dementia
Zuliani, G et al. "Relationship between low levels of high-density lipoprotein cholesterol and dementia in the elderly. The InChianti study". *Journals of Gerontology. Series A, Biological Sciences and Medical Sciences.* 2010 May;65(5):559-64

The object of this University of Ferrara study was to evaluate the association between cholesterol levels and the prevalence of dementia. At the start of the study, a total of 1,051 individuals aged 65 years or over were assessed for dementia. They were assessed again after three years.

The study found:

- After one year, individuals with dementia had significantly lower cholesterol levels and significantly lower levels of high-density lipoprotein (HDL) cholesterol compared with individuals without dementia.

- After three years, individuals newly diagnosed with dementia had significantly lower cholesterol levels and significantly lower levels of high-density lipoprotein (HDL) cholesterol compared with individuals without dementia.

The results of this Italian study demonstrate that low cholesterol levels and low levels of HDL cholesterol are associated with higher rates of dementia.

To summarise the many deleterious effects of low cholesterol on health as shown in this chapter, please follow the life story of someone I'll call Alex and some of the people that were involved in Alex's life.

Alex's story

Alex's mother was back on her feet within a couple of days of Alex's birth, and she was absolutely brimming with energy and happiness. Friends and family were amazed at how quickly she recovered from the pregnancy and astounded at her stamina and joyous demeanour.

Whilst she was in hospital with Alex, Alex's mother had a number of medical checks undertaken, including a measurement of her cholesterol. The doctor mentioned it seemed "a little high". Alex's mum had never taken any notice of "cholesterol", (she just lived her life), and thought nothing of it.

Alex's mother had a friend who was expecting a baby at the same time. Baby Morgan was born on the same day as Alex.

LOW CHOLESTEROL LEADS TO AN EARLY DEATH

The pregnancy had taken its toll on Morgan's mother, and for months after the birth, she was totally fatigued and had a severe depression.

She also had some tests performed on her in hospital, including a cholesterol check. The doctor had told her she should be "pleased with herself" as her cholesterol levels were very low.

Why might cholesterol levels be important regarding energy levels and depression in Alex and Morgan's mothers?

Cholesterol is the precursor to the adrenal hormones cortisol, corticosterone, aldosterone, and androstenedione and low cholesterol levels may disrupt the production of these essential hormones. This may result in "adrenal burnout," with symptoms of constant fatigue *(see paper 100)*.

Low serotonin levels are linked with depression. Serotonin is a neurotransmitter that helps elevate our mood. The body needs a certain amount of cholesterol circulating in its system to keep our neurons firing normally. So depleted cholesterol levels may impair the function of our nerve cells *(see paper 100)*.

Three other babies were born in the hospital that day. Baby Jamie was diagnosed with Smith-Lemli-Opitz syndrome and died within a week.

Baby Taylor and baby Jordan were also delivered.

Both Taylor and Jordan were later diagnosed as autistic by the age of three. They were badly behaved and were often irritable and hyperactive. Both were small for their age and found it very hard to interact and would not have eye contact. Having a massive temper tantrum and pulling their own hair out was a common occurrence.

In contrast, Alex was a lovely, contented toddler who enjoyed playing with older cousins and relished the activities at pre-school.

Within a couple of years, Taylor had stopped displaying any autistic behaviour and went on to lead a full, productive life, whereas Jordan's behaviour remained erratic and in adulthood, he could not hold down a regular job because of the bullying and discrimination he suffered as a result of his unusual behaviour.

What role do cholesterol levels have with Smith-Lemli-Opitz syndrome and autism?

Children born with Smith-Lemli-Opitz syndrome are unable to make enough cholesterol to support normal growth and development *(see paper 100)*, and as cholesterol is an essential component of the cell membrane and tissues of the brain, a person who can't make enough cholesterol will therefore experience poor growth, developmental delays, and mental retardation.

Children with severely low cholesterol will most likely be stillborn or die very early on in life. This was the outcome for baby Jamie.

Alex, Taylor and Jordan had their cholesterol levels measured at age three (around the same time as Taylor and Jordan were diagnosed as autistic). Alex apparently had "high cholesterol" whilst the parents of Taylor and Jordan were told their children had low cholesterol.

Cholesterol is crucial for the proper development and maintenance of the brain *(see paper 100)*. So low levels of cholesterol can lead to mental dysfunction, and in the cases of Taylor and Jordan, this manifested as autism.

Taylor was lucky. After he was diagnosed with autism, his parents fed him a high-cholesterol diet, and because cholesterol is essential for brain and physical development, Taylor experienced massive improvements in behaviour, social interactions and sleep patterns.

Alex entered the teenage years as a happy, confident, well-behaved, contented young person with many friends, and had never been in any trouble. Alex's cholesterol levels remained "high".

One of Alex's classmates had been Peter. Unfortunately, Peter had recently been expelled from school.

Peter's career at school had been blighted by bad behaviour and many suspensions. He was a very aggressive young man and frequently engaged in violent fights, which eventually contributed to his expulsion.

An analysis of his cholesterol showed it to be on the low side.

How do low cholesterol levels have a relationship with aggression and violence?

Low cholesterol is associated with a decrease in tryptophan. Tryptophan is the primary building block of the brain chemical serotonin. So low levels of tryptophan result in depleted serotonin levels.

As we have seen, low serotonin levels are associated with depression (as we saw with Morgan's mother). Low serotonin (caused by low cholesterol levels) is also strongly linked to suicide and impulsive aggression, which may be why Peter, with his low cholesterol levels, was prone to outbursts of aggression and violence.

Alex started work at the age of 22, after successful and rewarding years at university. On the opposite side of Alex's desk sat George, also 22, and also a new starter.

As part of the medical examination required for the job, Alex and George had to undergo a cholesterol test. The nurse was concerned about Alex's cholesterol reading, as it was elevated above the "norm", whereas George had an apparent "stellar" low cholesterol level.

After only a couple of weeks into the job, Alex noticed that George exhibited rather bizarre behaviour. George was paranoid about everyone in the office talking about her, was totally deluded regarding her role in the company, was completely disorganised and kept "hearing voices".

George was soon fired, and Alex later heard that George had been diagnosed with schizophrenia.

Conversely, Alex loved the job and soon gained a few promotions.

Again, low cholesterol levels have adverse effects on serotonin function, which could have impacted George's mental state.

As the journey of life continued, Alex had many adventures, made a wide circle of friends and fully participated in "life".

Alex's company had a "healthy awareness day", where health practitioners visited the office, conducted medical tests and gave nutritional advice. You've guessed it – they all had their cholesterol levels checked (as well as weight and blood pressure measurements). Alex still had "high cholesterol".

A pattern emerged in the workforce. The people with higher cholesterol tended to be more robust, healthier, rarely had a cough or cold and on their trips to exotic places, hardly ever contracted any illnesses. It was as if they were almost immune to illness.

On the other hand, those with lower cholesterol always seemed to suffer from sore throats and runny noses. Abdominal pain, vomiting, diarrhoea were common, and many had stays in hospital with an infectious disease.

Why do those with high cholesterol have better health?

People with higher cholesterol have a more robust immune system. Their bodies contain the resources to fight infections, whereas those with low cholesterol are deficient in the mechanisms that keep the body healthy *(see paper 101)*.

After reaching the age of 30, Alex settled down, married and started a family. The years started to roll by and Alex developed a love of the mountains and engaged in hobbies such as hiking and skiing. Alex's growing family enjoyed many holidays in the mountains with a large group of friends.

Medical check-ups were compulsory at Alex's office and every year Alex was always warned of the dangers of "high cholesterol".

Over the years, the large group of friends on the skiing holidays slowly dwindled as a result of ill health. Some of the friends struggled to walk very far and were diagnosed with conditions such as intermittent claudication and gout. However, Alex had no such trouble and immensely enjoyed the sport.

Almost invariably, those who were not well enough to partake in the holidays had lower cholesterol levels.

Low cholesterol is associated with increased levels of intermittent claudication *(see paper 83)* and increased levels of gout *(see paper 84)*.

50 years and more had now passed in Alex's life. The career was still moving upwards and the children were starting to become more independent.

Alex still loved learning and trying new things. Engaging in lively debate over a wide range of subjects was exciting and stimulating. Even general knowledge quiz games were played with gusto. Oh, cholesterol levels were still considered "high".

A number of Alex's friends did not share the enthusiasm for new challenges. Apathy had started to set in, concentration spans were short and memories were beginning to fail. Typical of this malaise was Sam.

Sam had been to the doctor to "find out what was up". A battery of tests was conducted and nothing was found. In fact, the doctor congratulated Sam on the very low cholesterol levels achieved.

How does low cholesterol affect learning and cognition?

About 40-60% of the brain is cholesterol, and it is a major component of the protective substance covering nerve cells (myelin). It also plays a major role in the development, function and stability of synapses (the connections between nerve cells – see paper 100).

It's no wonder that Sam feels apathetic and "can't be bothered". Sam's brain is starved of the raw material (cholesterol) that is needed to operate fully functionally.

Of course, Alex's brain and nerve cells are bathed in a rich cholesterol mix. This enables Alex to function at a high level.

Alex's life now moves serenely through the eighth, ninth and tenth decades of life. Despite being in the "golden years", Alex still regularly hikes up mountains and sometimes skis down them. Alex now enjoys the company of grandchildren and fully partakes in family matters and occasions. Alex still enjoys robust good health, a lively brain, and a fantastic memory – and doesn't even have to wear glasses. Alex's cholesterol is now higher than ever and most of the doctors who told Alex about the dangers of "high cholesterol" died long ago.

Some of Alex's friends also reached old age. Some died early.

Many of those friends started to succumb to "old-age" diseases from their fifties and onwards. The debilitating effects of Parkinson's, Alzheimer's and dementia reared their heads for many. It may start with a shuffling gait and mild confusion. Irritability, mood swings and aggression take over. Failing memory and befuddlement then take hold. It ends in total disorientation and helplessness. Many then require around the clock care.

What contributes to this degeneration in people's bodies and minds that leads to Parkinson's, Alzheimer's and dementia?

What helps Alex's brain to fully function at an advanced age?

Although the brain is only about 2% of body weight, it contains about a quarter of the total cholesterol in the body. This fact alone indicates how important cholesterol is to the brain.

Cholesterol is required everywhere in the brain as an antioxidant and helps rid the body of environmental toxins that might trigger Parkinson's.

LOW CHOLESTEROL LEADS TO AN EARLY DEATH

Cholesterol is a precursor for the hormones that are important in the central nervous system function.

Cholesterol makes up half of the volume of cell membranes.

Cholesterol and fat make up 75% of the protective sheath (myelin) covering nerve cells.

Cholesterol is essential for the formation of the links (synapses) between neurons (cells that carry messages to the brain).

I hope Alex's life story helps to clarify the vital importance of cholesterol for the healthy functioning of the human body.

Despite the scientific evidence showing that having low cholesterol will probably mean you will die earlier, a visit to your doctor for a medical check-up will result in a measurement of your cholesterol levels, and then in an almost reflex action response, you will be prescribed some statin drugs to lower your cholesterol and you will be advised to embrace a low-fat diet.

Chapter six looks at the scientific evidence to see the effects of drugs and diets on health and death rates.

CHAPTER 6

The effects of drugs and diet

The relentless drive to convert the population to a lifetime of cholesterol-lowering prescription drug usage, and the wholesale consumption of heavily processed, low-fat, low-cholesterol food products continues unabated.

Is it wise to heed these seemingly unstoppable dogmas?

Should we all be lining up at the pharmacy door to collect medications that reduce our cholesterol?

Should we ignore real food, and instead only buy products that are only available in a tin, packet or box that contain no fat and no cholesterol, and that have a long list of ingredients that are unpronounceable?

This chapter investigates the health consequences of complying with the ingestion of cholesterol-lowering drugs and adopting a low fat diet.

Paper 90
Statins increase the death rate by 150-300%
Bradford, RH et al. "Expanded Clinical Evaluation of Lovastatin (EXCEL) Study Results: 1. Efficacy in Modifying Plasma Lipoproteins and Adverse Event

Profile in 8245 Patients With Moderate Hypercholesterolemia". *Archives of Internal Medicine*. 1991; 151 (1): 43-49

In this study, Dr Reagan Bradford and his team investigated the relationship between statins and death rates. In the study, 8,245 "patients", aged 18 to 70, with cholesterol levels between 232mg/dL (6.0 mmol/L) and 300mg/dL (7.8 mmol/L) received one of four different doses of lovastatin (Mevacor) or a placebo.

The study found after one year:

- Higher transaminase levels (which may be an indicator of liver damage) were found in the subjects taking statins.
- Higher incidence of clinical adverse experiences requiring patients to discontinue the "treatment" was found in the subjects taking statins.
  - 16% more patients taking 20 mg/day statins discontinued their treatment compared to patients taking a placebo.
  - 50% more patients taking 80 mg/day statins discontinued their treatment compared to patients taking a placebo.
- Higher levels of muscle damage were detected in the subjects taking statins.
- The four groups taking lovastatin lowered their low-density lipoprotein (LDL) cholesterol levels by 24%-40%.
- The four groups taking lovastatin lowered their cholesterol levels by 17%-29%.

- The death rate of the four groups taking various doses of lovastatin was 150-300% higher than the placebo group.

The results of the study show that statin drugs lower cholesterol levels and increase the death rate.

Paper 91
The elderly die earlier when taking statins

Petersen, LK et al. "Lipid-lowering treatment to the end? A review of observational studies and RCTs on cholesterol and mortality in 80+-year olds". *Age and Ageing.* 2010 39 (6): 674-680

The lead author of the study, Dr Line Kirkeby Petersen, is a researcher at the Danish Aging Research Centre. Dr Petersen reviewed a total of 12 studies regarding the relationship of cholesterol levels and all-cause mortality. This corresponded to 13,622 participants: 3,789 aged 80 and above in eight studies and 9,833 aged 71–103 (average age 78 years) in four studies.

The review found:

- Regarding all-cause mortality in the 80+-year olds, low cholesterol levels did not seem beneficial in any study and low cholesterol levels (less than 5.5 mmol/l or 212 mg/dL) is associated with increased mortality among 80+-year olds.
- There was no evidence that statins decrease all-cause mortality in elderly people without known vascular disease, on the contrary, it was even possible that statins increased all-cause mortality.

This finding of the review show that low cholesterol and possibly statin treatment increases the death rate in the elderly.

Paper 92
Men treated with cholesterol lowering drugs have a 142% increase in heart disease deaths

Strandberg, TE et al. "Long-term Mortality After 5-Year Multifactorial Primary Prevention of Cardiovascular Diseases in Middle-aged Men". *Journal of the American Medical Association.* 1991;266:1225-1229

This trial lasted 15 years (it included 1,222 men who had heart disease risk factors) and investigated the effects of various preventative measures on total death rate and heart disease death rate.

The men were split into two groups:

- Group (a) (intervention group) visited the investigators every fourth month. They were treated with intensive dietetic-hygienic measures and frequently with cholesterol lowering drugs (mainly clofibrate and/or probucol) and blood pressure drugs (mainly beta-blockers and/or diuretics).
- Group (b) (control group) were not treated by the investigators.

After 15 years, the study revealed:

- There was a 45% increase in deaths in the intervention group compared to the control group.
- There was a 142% increase in heart disease deaths in the intervention group compared to the control group.

- There was a 1,300% increase in deaths due to violence in the intervention group compared to the control group.

The results from this trial showed that men who followed the cholesterol-lowering regime were more likely to die earlier and more than twice as likely to die of heart disease.

Paper 93
The lower your cholesterol is, the more likely you are to die

Kame, C et al. "Estimation of Effect of Lipid Lowering Treatment on Total Mortality Rate and Its Cost-Effectiveness Determined by Intervention Study of Hypercholesterolemia". *Japanese Journal of Hygiene.* Vol. 62 (2007), No. 1 p.39-46

This study analysed data from 55,000 people aged between 30 and 69 and estimated the effects on death rates assuming that cholesterol levels decrease from 240-259 mg/dL (6.24-6.73 mmol/L) to 160-179 mg/dL (4.2-4.6 mmol/L) by use of drug therapy.

The data revealed:

- In men, cholesterol levels falling from 240-259 mg/dL (6.24-6.73 mmol/L) to 160-179 mg/dL (4.2-4.6 mmol/L) would increase the death rate by 145%.
- In women, cholesterol levels falling from 240-259 mg/dL (6.24-6.73 mmol/L) to 160-179 mg/dL (4.2-4.6 mmol/L) would increase the death rate by 154%.

The study showed that lowering cholesterol levels with drug therapy results in about a 150% higher death rate.

Paper 94
Lower cholesterol levels result in higher cardiac death rates and higher total death rates

Frick, MH et al. "Efficacy of gemfibrozil in dyslipidaemic subjects with suspected heart disease. An ancillary study in the Helsinki Heart Study frame population". *Annals of Medicine*. 1993 Feb;25(1):41-5

This study was conducted by Dr M Hekki Frick from Helsinki University Central Hospital. The study examined the effects of gemfibrozil (a cholesterol-lowering drug) on death rates in male patients who had symptoms and signs of possible coronary heart disease. The study included 626 patients, with an average age of 49, who were followed for five years.

The men were given either:

- 600 mg gemfibrozil twice daily.
- Placebo.

Dr Frick discovered:

- The men taking gemfibrozil lowered their cholesterol levels by 8.5%.
- The men taking gemfibrozil had 61% more deaths than the men taking a placebo.
- The men taking gemfibrozil had 117% more cardiac deaths than the men taking a placebo.

The results of the study show that taking the cholesterol-lowering drug gemfibrozil results in lower cholesterol levels and higher cardiac death rates and higher total death rates.

Paper 95
The adverse health effects of low cholesterol
Song, JX et al. "Primary and secondary hypocholesterolemia". *Journal of Peking University.* 2010 Oct 18;42(5):612-5

In this review of the literature, Dr Jun-xian Song from the Department of Cardiology, Peking University examined the influence of low cholesterol levels (hypocholesterolemia) on health.

The review found:

- Low cholesterol levels are common in the population.
- Physicians pay little attention to the diseases, causes and consequences of low cholesterol in clinical practice.
- Low cholesterol levels can result in some adverse events, such as increased death rates, intracerebral hemorrhage, cancer, infection, adrenal failure, suicide and mental disorder.
- Despite the adverse health consequences of low cholesterol, physicians are increasingly prescribing cholesterol-lowering treatments such as statin drugs.

With all the adverse health effects of low cholesterol, Dr Song concludes: *"It's high time that physicians attached more importance to hypocholesterolemia."*

Paper 96
Low fat, low cholesterol diets result in lower cholesterol levels and a higher death rate

Woodhill, JM et al. "Low fat, low cholesterol diet in secondary prevention of coronary heart disease". *Advances in Experimental Medicine and Biology.* 1978;109:317-30

This study set out to determine the effects of a low fat, low cholesterol diet on death rates in men with existing heart disease. The study included 458 men, aged 30 to 59, who were followed for up to seven years.

The men were allocated in groups to consume either:

- A low fat, low cholesterol diet.
- Their usual diet.

The study revealed:

- The men following the low-fat, low-cholesterol diet lowered their cholesterol levels 4.5% more than the men following their usual diet.
- The men following the low-fat, low-cholesterol diet had 49% increased death rates compared to the men following their usual diet.
- The men following the low-fat, low-cholesterol diet had 44% increased heart disease death rates compared to the men following their usual diet.

The results of the study show that a low-fat, low-cholesterol diet results in lower cholesterol levels and a higher death rate.

Paper 97
Low-cholesterol, low saturated fat diets lead to a fall in cholesterol levels and an increase in heart attacks and death rates

Frantz, ID et al. "Test of effect of lipid lowering by diet on cardiovascular risk. The Minnesota Coronary Survey". *Arteriosclerosis*. 1989 Jan-Feb;9(1):129-35

This University of Minnesota study compared the effects of two diets on cholesterol levels and the incidence of heart attacks, sudden deaths, and all-cause mortality. The trial included 9,057 men and women, who were followed for an average of just over a year.

The diets were either:

- 39% fat control diet (18% saturated fat, 5% polyunsaturated fat, 16% monounsaturated fat, 446 mg dietary cholesterol per day) (High saturated fat, high-cholesterol diet)
- 38% fat treatment diet (9% saturated fat, 15% polyunsaturated fat, 14% monounsaturated fat, 166 mg dietary cholesterol per day) (Low saturated fat, low-cholesterol diet)

The researchers found:

- Cholesterol levels remained similar on the high saturated fat, high-cholesterol diet.
- Cholesterol levels fell by 16% on the low saturated fat, low-cholesterol diet.
- Those on the low saturated fat, low cholesterol diet had a 5% increased risk of heart attack and sudden

death compared to those on the high saturated fat, high-cholesterol diet.
- Those on the low saturated fat, low cholesterol diet had a 6% increase in death rates compared to those on the high saturated fat, high-cholesterol diet.

This study reveals that low-cholesterol, low saturated fat diets lead to a fall in cholesterol levels and an increase in heart attacks and death rates.

Paper 98
Professor finds the best way to raise HDL (good cholesterol) is to eat saturated fat

Mensink, RP et al. "Effect of dietary fatty acids on serum lipids and lipoproteins. A meta- analysis of 27 trials". *Arteriosclerosis and Thrombosis.* Vol 12, 911-919

Professor Ronald Mensink and his team from Limburg University, Maastricht, reviewed 27 studies to calculate the effect of replacing carbohydrates with either saturated fat, monounsaturated fat or polyunsaturated fat on (the beneficial) high-density lipoprotein (HDL) cholesterol levels and (the harmful) triglyceride levels.

This review of 27 studies revealed:

- All fats lowered the harmful triglyceride levels.
- All fats raised the beneficial HDL cholesterol levels, but saturated fat raised it the most!

Professor Mensink found the best way to raise HDL (good cholesterol) is to eat saturated fat.

HDL cholesterol is linked to a longer life expectancy *(see papers 28-32).*

Paper 99
Lowering cholesterol results in an increase in death rates

Atrens, DM. "The questionable wisdom of a low-fat diet and cholesterol reduction". *Social Science and Medicine.* Volume 39, Issue 3, August 1994, Pages 433-447

Dr Dale Atrens from the University of Sydney reviewed the scientific literature regarding low-fat diets, cholesterol lowering, heart disease and health.

Dr Atrens discovered:

- An examination of the foundations that a low-fat diet and cholesterol reduction are essential to good cardiovascular health suggests that in many respects it was ill-conceived from the outset and, with the accumulation of new evidence, it is becoming progressively less tenable.
- Many studies have suggested that increased dietary fat intake may decrease death from heart disease.
- High cholesterol is frequently associated with increased overall life expectancy.
- Numerous studies have shown that lifestyle and dietary advice to lower cholesterol levels may increase death from cardiovascular causes.
- The only significant overall effect of cholesterol lowering that has ever been shown is increased death rates.

This review found that low-fat diets and cholesterol lowering do not lower heart disease rates. However,

LOW CHOLESTEROL LEADS TO AN EARLY DEATH

actively lowering cholesterol values will result in an increase in death rates.

Rather than helping us in our quest for health, cholesterol-lowering drugs and low-fat diets are shown to be disastrous to our well-being.

Statins and fibrate drugs both lower cholesterol levels, but also cause a plethora of side effects and an increase in many types of death.

Not only are low-fat diets insipid and tasteless, they can also result in higher death rates from heart disease, whereas a high saturated-fat diet may be linked to a higher life expectancy.

To quote Dr Dale Atrens *"The only significant overall effect of cholesterol-lowering that has ever been shown is increased death rates"*.

Chapter seven includes a couple of papers that show the advantageous properties of cholesterol.

CHAPTER 7

How does cholesterol help us to live longer?

The following two studies help to illustrate the many roles that cholesterol plays in helping us to maintain good health, which helps to explain why cholesterol is vital for life and helps us to live longer.

Paper 100
Cholesterol supplementation benefits patients with Smith-Lemli-Opitz syndrome
Elias, ER et al. "Clinical effects of cholesterol supplementation in six patients with Smith-Lemli-Opitz syndrome (SLOS)".
American Journal of Medical Genetics.
1997 Jan 31;68(3):305-10

Children with Smith-Lemli-Opitz syndrome have very low cholesterol levels and most are either stillborn or die early because of serious malformations of the central nervous system. Those that survive have a small head size, learning problems and behavioural problems. They tend to grow more slowly than other infants and many affected individuals have fused second and third toes and some have extra fingers or toes.

LOW CHOLESTEROL LEADS TO AN EARLY DEATH

The problems that occur in Smith-Lemli-Opitz syndrome are because of their very low cholesterol levels, since cholesterol is probably the most important molecule in the body.

Cholesterol plays many vital roles:

- It is necessary for normal embryonic development and has important functions both before and after birth.
- It is a structural component of cell membranes.
- About 40-60% of the brain is cholesterol.
- It is a major component of the protective substance (myelin) covering nerve cells.
- It plays a major role in the development, function and stability of synapses (the connections between nerve cells).
- It plays a role in the production of vitamin D, certain hormones and bile acids.
- It is needed to produce vital hormones such as testosterone, progesterone and estrogen.
- Cholesterol is the precursor to the adrenal hormones cortisol, corticosterone, aldosterone, and androstenedione.
- It is essential for the production of bile acids that are vital for the absorption of certain minerals and the fat-soluble vitamins A, D, E and K.
- It helps to *repair* damaged arteries.

This study, based at the Department of Pediatrics, Tufts-New England Medical Center, Boston, examined the effects of cholesterol supplementation in children with

Smith-Lemli-Opitz syndrome. The trial included six children, ranging in age from birth to 11 years old.

The researchers found:

- Clinical benefits of the cholesterol therapy were seen in all patients, irrespective of their age at the onset of treatment, or the severity of their cholesterol defect.
- The cholesterol therapy improved growth, gave a more rapid developmental progress, and a lessening of problem behaviours. As older patients progressed to puberty, they had a better tolerance of infection, improvement of gastrointestinal symptoms, and a reduction in photosensitivity and skin rashes.
- Patients had no adverse reactions to treatment with cholesterol.

The results of the study show that cholesterol supplementation benefits patients with Smith-Lemli-Opitz syndrome.

Paper 101
High cholesterol levels strengthen the immune system

Muldoon, MF et al. "Immune system differences in men with hypo- or hypercholesterolemia".
Clinical Immunology and Immunopathology.
1997 Aug;84(2):145-9

Lymphocytes are a type of white blood cell in the blood. White blood cells help to protect the body against diseases and infections.

T cells are a type of white blood that directs the body's immune system to defend against bacteria and other harmful cells.

CD8+ cells are T cells that can kill cells that are infected with viruses, other pathogens, or are otherwise damaged or dysfunctional.

CD4+ cells are T cells that release cytokines (cytokines are substances that communicate to the immune system that there are "foreign harmful invaders").

Interleukin-2 is a cytokine that responds to the poison phytohemagglutinin. Phytohemagglutinin is found in large quantities in red kidney beans and in lesser amounts in white kidney beans, fava beans and other leguminous beans.

Professor Matthew Muldoon from the University of Pittsburgh notes that low cholesterol levels in apparently healthy individuals are associated with increased death rates from cancer and other causes of death.

This study examined the immune systems in men with low cholesterol and in men with high cholesterol to find what may explain these associations.

The study found:

- Men with low cholesterol levels, below 151 mg/dL (3.9 mmol/L), had significantly fewer lymphocytes than men with high cholesterol levels, above 261 mg/dL (6.7 mmol/L).

- Men with low cholesterol levels, below 151 mg/dL (3.9 mmol/L), had fewer T cells than men with high cholesterol levels, above 261 mg/dL (6.7 mmol/L).
- Men with low cholesterol levels, below 151 mg/dL (3.9 mmol/L), had fewer CD8+ cells than men with high cholesterol levels, above 261 mg/dL (6.7 mmol/L).
- Men with low cholesterol levels, below 151 mg/dL (3.9 mmol/L), had fewer CD4+ cells than men with high cholesterol levels, above 261 mg/dL (6.7 mmol/L).
- Men with low cholesterol levels, below 151 mg/dL (3.9 mmol/L), had less interleukin-2 release in response to phytohemagglutinin than men with high cholesterol levels, above 261 mg/dL (6.7 mmol/L).

The data from this study shows that men with high cholesterol levels have a stronger immune system than men with low cholesterol levels.

This may help to explain why people with high cholesterol live longer than people with low cholesterol.

CHAPTER 8

Summary of the evidence

The scientific evidence has revealed that it is not a good idea to lower your cholesterol levels, because men and women with low cholesterol die earlier.

Falling cholesterol levels also lead to a reduced lifespan, and low cholesterol levels correlate with more deaths in people that have been hospitalised.

Low levels of the beneficial high-density lipoprotein (HDL) cholesterol and even low levels of the so-called "bad" low-density lipoprotein (LDL) cholesterol lead to higher death rates.

The "high cholesterol causes heart disease" dogma is shown to be a complete myth. Statistics collated from around the world actually show that as cholesterol levels increase, then deaths from cardiovascular diseases decrease.

Low cholesterol levels are implicated in an increase in the rates of ill health and death in a plethora of conditions and diseases.

Cholesterol lowering drugs, such as statins, and low-fat diets are not only ineffective, but have many side effects that may decrease our quality of health and shorten our lifespan.

It is revealed that cholesterol is a vital, necessary and essential substance that our body needs for us to live a long healthy life.

This review of the evidence demonstrates:

- Low cholesterol levels are associated with a shorter life.
- Low levels of HDL and LDL cholesterol are linked to a shorter lifespan.
- High cholesterol does *not* cause heart disease.
- Low cholesterol leads to illness and death in many diseases and conditions.
- Statin drugs and low-fat diets may lead to higher death rates.
- Saturated fat can give protection from heart disease.
- Cholesterol is an essential substance needed for a long, healthy life.

Appendix 1

Glossary

Adrenal failure
A condition in which the adrenal glands do not produce enough of the adrenal hormones.

Adrenal glands
These are located on top of the kidneys. They are chiefly responsible for releasing hormones in response to stress.

Adrenal hormones
Hormones such as cortisol, corticosterone, aldosterone, and androstenedione.

Albumin
A protein in your bloodstream that helps transport a variety of important substances, including calcium, hormones, the protein bilirubin and important nutrients called fatty acids. Albumin also helps your blood maintain its osmotic pressure, which helps keep its water content from leaking through your blood vessels into surrounding tissue.

Aldosterone
A hormone produced by the adrenal gland. Aldosterone regulates the balance of water and electrolytes in the body.

Alzheimer's
A terminal, progressive disease of the brain, and the most common form of dementia.

Amino acids
The building blocks of proteins.

Anaemia
A deficiency in the number of red blood cells or in their haemoglobin content, resulting in pallor, shortness of breath, and lack of energy.

Androstenedione
A steroid hormone produced in the adrenal gland that is a precursor to testosterone and other male hormones (androgens).

Angina
Chest pain or discomfort that occurs if an area of your heart muscle doesn't get enough oxygen-rich blood.

Antigen
Any substance that causes your immune system to produce antibodies against it. An antigen may be a foreign substance from the environment such as chemicals, bacteria, viruses, or pollen. An antigen may also be formed within the body, as with bacterial toxins or tissue cells.

Antioxidants
Substances that may protect cells from the damage caused by unstable molecules known as free radicals.

Anti-social personality disorder
A mental health condition in which a person has a long-term pattern of manipulating, exploiting, or violating the rights of others.

Apolipoproteins
Proteins that bind fat and cholesterol to form lipoproteins.

Appendicitis
Swelling (inflammation) of the appendix.

Arthropathy
Disease of the joints.

Asbestosis
Serious long-term lung condition caused by breathing in asbestos dust.

Asthma
Common inflammation of the airways.

Autism
A range of complex neurodevelopment disorders, characterised by social impairments, communication difficulties, and restricted, repetitive, and stereotyped patterns of behaviour.

Bacteremia
The presence of bacteria in the blood.

Bartholin cyst or abscess
An infection of the Bartholin's glands, which lie next to the entrance to the vagina.

Beta-blockers
A class of drug used for treating abnormal heart rhythms and high blood pressure.

Bile acids
Vital for the absorption of certain minerals and the fat-soluble vitamins A, D, E and K.

Boils
Skin infections that start in a hair follicle or oil gland.

Bronchitis
Inflammation of the main air passages to the lungs.

Carbuncle
An abscess larger than a boil, usually with one or more openings draining pus onto the skin. It is usually caused by bacterial infection.

Cardiovascular diseases
A class of diseases that involve the heart or blood vessels, such as heart disease, heart failure and stroke.

CD4+ cells
T cells that release cytokines.

CD8+ cells
T cells that can kill cells that are infected with viruses, other pathogens, or are otherwise damaged or dysfunctional.

Cellulitis
An infection of the skin and the tissues just below the skin surface.

Central nervous system
Consists of the brain and spinal cord. It contains millions of neurones (nerve cells).

Cerebrovascular system
Pertaining to the blood vessels of the cerebrum, or brain.

Cervicitis
An inflammation (irritation) of the lining of the cervix.

Chlamydia
A common sexually transmitted infection (STI) in humans caused by the bacterium Chlamydia trachomatis.

Cholangitis
An infection of the common bile duct, the tube that carries bile from the liver to the gallbladder and intestines.

Cholesterol
A waxy substance found in your body that is needed to produce hormones, vitamin D and bile. Cholesterol is also important for protecting nerves and for the structure of cells.

Chronic obstructive pulmonary disease (COPD)
The occurrence of chronic bronchitis or emphysema, a pair of commonly co-existing diseases of the lungs in which the airways become narrowed.

Cognition
Refers to mental processes such as attention, memory, producing and understanding language, solving problems, and making decisions.

Clofibrate
A type of fibrate.

Conduct disorder
A disorder of childhood and adolescence that involves long-term behaviour problems.

Coronary artery disease
Another term for coronary heart disease

Coronary heart disease
The narrowing or blockage of the coronary arteries.

Cortisol
A steroid hormone produced by the adrenal gland. It is released in response to stress.

Corticosterone
A steroid hormone produced in the adrenal glands. It protects against stress.

Crohn's disease
A long-term gut condition that causes inflammation of the lining of the digestive system.

Cystitis
A common infection of the bladder.

Cytokines
Substances that communicate to the immune system that there are "foreign harmful invaders".

Dementia
A group of diseases that cause a permanent decline of person's ability to think, reason and manage their own life.

Dermatitis
Inflammation of the skin.

Digestive system
Consists of organs that break down food into components that your body uses for energy and for building and repairing cells and tissues.

Dilated cardiomyopathy
A condition in which the heart becomes weakened and enlarged and cannot pump blood efficiently.

Diuretics
Also known as water pills, are prescribed to lower blood pressure.

Diverticulosis
Refers to the presence of small out-pouchings (called diverticula) or sacs that can develop in the lining of the gastrointestinal tract. It may cause pain or discomfort in the left lower abdomen, bloating, and/or a change in bowel habits.

Eczema
A form of dermatitis, or inflammation of the epidermis (the outer layer of the skin).

Electrolytes
Substances found in the blood, such as sodium and potassium.

Empyema
A collection of pus in the space between the lung and the inner surface of the chest wall.

Emphysema
A chronic lung disease.

Endocarditis
Inflammation of the inside lining of the heart chambers and heart valves (endocardium).

Endotoxic shock
A serious condition that occurs when an overwhelming infection leads to life-threatening low blood pressure.

Encephalitis
Acute inflammation (swelling up) of the brain resulting either from a viral infection or when the body's own immune system mistakenly attacks brain tissue.

End-stage kidney disease
The complete or almost complete failure of the kidneys to work.

End-stage renal disease
Another term for end-stage kidney disease

Epididymitis
Swelling (inflammation) of the epididymis, the tube that connects the testicle with the vas deferens.

Firates
Type of drugs used to lower cholesterol levels.

Free radicals
Molecules responsible for aging and tissue damage.

Gangrene
The death of tissue in part of the body.

Gastroenteritis
An infection of the intestines which usually causes vomiting and diarrhoea.

Gastrointestinal
Refers to the stomach and intestine, and sometimes to all the structures from the mouth to the anus.

Gemfibrozil
A cholesterol-lowering drug.

Gonorrhoea
A sexually transmitted infection caused by bacteria called Neisseria gonorrhoeae or gonococcus.

Gout
A form of acute arthritis that causes severe pain and swelling in the joints.

Gynaecological
Referring to the female reproductive system.

Haemoglobin
A protein found in the red blood cells that carry oxygen around your body and gives blood its red colour.

Heart attack
Occurs if the flow of oxygen-rich blood to a section of heart muscle suddenly becomes blocked. If blood flow isn't restored quickly, the section of heart muscle begins to die.

Heart failure
When the heart loses its ability to pump blood efficiently through your body.

Hematocrit
The percentage of the blood volume occupied by red blood cells.

Hemodialysis
Where a dialysis machine and a special filter called an artificial kidney, or a dialyzer, are used to clean your blood.

Hemodynamics
A measurement of blood pressure and blood flow.

Hemorrhagic stroke
Occurs when a weakened blood vessel ruptures.

Hepatitis B
Irritation and swelling (inflammation) of the liver due to infection with the hepatitis B virus (HBV).

Hepatitis C
An infectious disease affecting primarily the liver, caused by the hepatitis C virus (HCV).

Herpes
A viral disease caused by both Herpes simplex virus type 1 (HSV-1 cold sores and whitlows (lesions) on fingers and hands, also half new cases of genital herpes) and type 2 (HSV-2 genital sores, also sometimes cold sores and whitlows).

High-density lipoprotein (HDL)
Collects cholesterol from the body's tissues and brings it back to the liver.

High-density lipoprotein 2 (HDL2)
Large type of HDL.

High-density lipoprotein 3 (HDL3)
Small type of HDL

Hispanics
Americans with origins in the Hispanic countries of Latin America or Spain.

Hypocholesterolemia
Low cholesterol levels.

Impetigo
A highly contagious bacterial infection of the surface layers of the skin, which causes sores and blisters.

Infectious myositis
An infection of skeletal muscle.

Influenza
A viral infection that affects mainly the nose, throat, bronchi and, occasionally, lungs.

Interleukin-2
A cytokine that responds to the poison phytohemagglutinin.

Intermittent claudication
Pain in the leg brought on by walking and is caused by poor blood flow to the muscles. It is intermittent because it only comes on with walking or exercise and goes away when you rest.

Intracerebral hemorrhage
Occurs when a diseased blood vessel within the brain bursts, allowing blood to leak inside the brain.

Iron-deficiency anaemia
A low red blood cell level caused by insufficient dietary intake and absorption of iron and/or iron loss from intestinal bleeding, parasitic infection, menstruation, etc.

Ischemic heart disease
A disease characterised by reduced blood supply of the heart muscle, usually due to coronary artery disease.

Ischemic stroke
Occurs as a result of an obstruction within a blood vessel supplying blood to the brain. It accounts for 87 percent of all stroke cases.

Laryngitis
An inflammation of the larynx.

Left ventricular ejection fraction (LVEF)
The measurement of how much blood is being pumped out of the left ventricle of the heart (the main pumping chamber) with each contraction.

Lipoprotein
A molecule containing protein, fat and cholesterol that circulates around the bloodstream.

Low-density lipoprotein (LDL)
Carries cholesterol from the liver to cells of the body.

Lung disease
Any disease or disorder that occurs in the lungs or that causes the lungs to not work properly.

Lymphadenitis
An infection of the lymph nodes.

Lymphocytes
A type of white blood cell in the blood. White blood cells help to protect the body against diseases and infections.

Mediastinitis
Swelling and irritation (inflammation) of the area between the lungs (mediastinum).

Meningitis
Inflammation of the protective membranes covering the brain and spinal cord.

Meningococcal sepsis
Where bacteria has invaded the bloodstream.

Mortality rate
A measure of the number of deaths

Myelin
An insulating layer that forms around nerves, including those in the brain and spinal cord. It is made up of protein and fatty substances.

Myelitis
A disease involving inflammation of the spinal cord, which disrupts central nervous system functions linking the brain and limbs.

Myocardial infarction
Commonly known as a heart attack, results from the interruption of blood supply to a part of the heart, causing heart cells to die.

Neurones
Nerve cells.

Neurosurgery
The medical specialty concerned with the prevention, diagnosis, treatment, and rehabilitation of disorders which affect any portion of the nervous system, including the brain, spinal cord, peripheral nerves, and extra-cranial cerebrovascular system.

New York Heart Association (NYHA) Functional Classification
Provides a simple way of classifying the extent of heart failure.

Nonischemic heart disease
A disease of the heart that lacks the associated coronary artery disease often found in other diseases of the heart. It's usually linked to a disease in one or more of the cardiac muscles, causing the heart to pump in an ineffective manner, thereby reducing the transport of blood, oxygen and other nutrients throughout the body.

Oophoritis
An inflammation of the ovaries.

Orchitis
Swelling (inflammation) of one or both of the testicles.

Osteomyelitis
Infection of the bone or bone marrow.

Parkinson's
A degenerative disease of the nervous system associated with trembling of the arms and legs, stiffness and rigidity of the muscles and slowness of movement.

Pelvic inflammatory disease
A bacterial infection of the female upper genital tract, including the womb, fallopian tubes and ovaries.

Periostitis
Inflammation of the periosteum, a layer of connective tissue that surrounds bone.

Photosensitivity
Allergy to the sun.

Phytohaemagglutinin
Toxic agent found in many species of beans, but it is in highest concentration in red kidney beans.

Pilonidal cysts
Cysts that form near the buttocks.

Placebo
An inactive substance administered to a patient usually to compare its effects with those of a real treatment.

Plant stanol esters
A patented ingredient of Raisio Group that blocks the absorption of cholesterol and other important nutrients. The active ingredient of Benecol products.

Plant sterols
Food additive that blocks the absorption of cholesterol and other important nutrients.

Pleurisy
Inflammation of the lining of the lungs and chest (the pleura) that leads to chest pain when you take a breath or cough.

Pneumonia
An inflammation of the lung tissue affecting one or both sides of the chest that often occurs as a result of an infection.

Pneumoconiosis
An occupational lung disease and a restrictive lung disease caused by the inhalation of dust, often in mines

Probucol
Drug used to lower cholesterol.

Prostatitis
An inflammation of the prostate gland, in men.

Pulmonary congestion
A condition in which there is excess fluid built up in the air sacs of the lungs.

Pulmonary fibrosis
The formation or development of excess fibrous connective tissue (fibrosis) in the lungs.

Pyoderma
Skin infection that exudes pus.

Respiratory diseases
Diseases of the respiratory system such as pneumonia, bronchitis and emphysema.

Respiratory system
Supplies the blood with oxygen in order for the blood to deliver oxygen to all parts of the body. The respiratory system does this through breathing.

Reverse epidemiology
A term that suggests that obesity and high cholesterol may be protective and associated with greater survival in disease.

Rheumatic pneumonia
Pneumonia rarely occurring in severe acute rheumatic fever.

Rheumatoid arthritis
Causes inflammation, pain, and swelling of joints.

Rhinitis
Irritation and inflammation of the mucous membrane inside the nose.

Ringworm
A skin infection due to a fungus.

Rotavirus
A viral infection that can cause gastroenteritis.

Salmonella
A group of bacteria that can cause food poisoning. Typically, food poisoning causes gastroenteritis, an

infection of the gut (intestines) leading to diarrhoea and vomiting.

Salpingitis
An infection and inflammation in the fallopian tubes.

Saturated fat
A fat that has its full complement of hydrogen atoms. A very stable fat that is good for cooking with.

Scandinavian Stroke Scale
A measurement of stroke severity where a score of 0 is the most severe stroke and a score of 58 is the least severe stroke.

Schizophrenia
A mental illness. Symptoms include hallucinations (such as hearing voices), delusions (false ideas), disordered thoughts, and problems with feelings, behaviour and motivation.

Septicemia
The presence of pathogenic organisms in the bloodstream.

Septic shock
See endotoxic shock.

Serotonin
A compound that occurs in the brain, intestines, and blood platelets and transmits nerve impulses.

Sinusitis
An inflammation of the lining of one or more of the sinuses. These are air-filled cavities in the bones of the

skull and face, which connect with the nose through small openings. There are four pairs of sinuses: the frontal sinuses sit above the eyes in the forehead, the maxillary sinuses lie behind the cheekbones, the sphenoid pair rests behind the nose, and the ethmoid sinuses are located between the eyes and the bridge of the nose.

Smith-Lemli-Opitz syndrome (SLOS)
An inherited metabolic disorder in which cholesterol is not synthesised properly in the body.

Statins
A class of drugs used to lower cholesterol levels. They have many adverse side effects.

Stroke
The sudden death of brain cells in a localised area due to inadequate blood flow.

Subcutaneous tissue
The deepest layer of human skin. It is responsible for regulating body temperature and contains elastic fibres, nerves, and hair follicle roots.

Sudden cardiac death
The unexpected natural death from a cardiac (heart) cause within a short time period, generally less than one hour from the onset of symptoms, in a person without any prior condition that would appear fatal.

Surface antigen
The first marker to appear after hepatitis B virus infection, preceding clinical disease by weeks, peaking

with the onset of symptoms and disappearing six months post-infection.

Synapses
Structures that permit a neuron to pass an electrical or chemical signal to another cell (neural or otherwise).

Syphilis
A sexually transmitted disease caused by bacteria. It infects the genital area, lips, mouth, or anus of both men and women.

Systemic inflammation
Occurs when chronic inflammation moves beyond local tissues and into the lining of blood vessels and organs.

T cells
A type of white blood that directs the body's immune system to defend against bacteria and other harmful cells.

Thrush
A fungal infection of any of the *Candida* species (all yeasts), of which *Candida albicans* is the most common.

Tonsillitis
Inflammation of the tonsils, most commonly caused by a viral or bacterial infection.

Transaminase
An enzyme that catalyses chemical reactions. Elevated levels of transaminase may be an indicator of liver damage.

Trichomoniasis
A common sexually transmitted infection caused by a tiny parasite called Trichomonas vaginalis.

Trichotillomania
The compulsion to tear or pluck out the hair on one's head and face and often to ingest it.

Triglycerides
A type of fat in the bloodstream and fat tissue.

Tryptophan
An amino acid needed for normal growth in infants and for nitrogen balance in adults.

Tuberculosis (TB)
Caused by bacteria (*Mycobacterium tuberculosis*) that most often affect the lungs.

Ulcerative colitis (UC)
A disease where inflammation develops in the large intestine (the colon and rectum). The most common symptom when the disease flares up is diarrhoea mixed with blood.

Viral hepatitis
Inflammation of the liver caused by a virus. Several different viruses, named the hepatitis A, B, C, D, and E viruses, cause viral hepatitis.

Vaginitis
An inflammation of the vagina.

Vascular disease
A form of cardiovascular disease primarily affecting the blood vessels.

Vas deferens
Thick-walled tube in the male reproductive system that transports sperm cells from the epididymis, where the sperm are stored prior to ejaculation.

Appendix 2

Further resources

Websites:
www.thincs.org

The International Network of Cholesterol Skeptics (or THINCS) is a group of scientists, physicians, and other academicians from around the world who dispute the widely accepted saturated fat/cholesterol causes heart disease hypothesis. THINCS was founded in January 2003.

www.dietsandscience.com

I've run this website since 2010. I've written easy to read reviews on over 1,000 diet, lifestyle & health studies from research centres, universities and peer-reviewed journals.

Books:
Ignore the Awkward: How the Cholesterol Myths Are Kept Alive by Uffe Ravnskov

The Great Cholesterol Con by Malcolm Kendrick

Cholesterol and Saturated Fat Prevent Heart Disease by David Evans

The Great Cholesterol Con by Anthony Colpo

Fat and Cholesterol are Good For You by Uffe Ravnskov

Nutrition and Physical Degeneration by Weston A Price

DVDs:
Fat Head by Tom Naughton

Fat Head contradicts everything you've ever been told about diet and heart disease with true science to back it up. Themes include: having low cholesterol is unhealthy. Low-fat diets can lead to depression and type II diabetes. Saturated fat doesn't cause heart disease — but sugars, starches and processed vegetable oils do.

In Search of the Perfect Human Diet CJ Hunt

Researchers, scientists, and nutritionists discuss why the original human diet is still the diet we should be eating today.

Appendix 3

List of studies

Chapter 1: The lower your cholesterol, the earlier you die

1. Low cholesterol is associated with increased death rates in 35-74 year olds
2. Study of 482,472 men aged 30-65 shows that low cholesterol increases the mortality rate by 35% in men
3. High cholesterol results in a 13% lower death rate compared to low cholesterol in 35-64 year olds
4. Higher cholesterol levels associated with lower death rates in 350,977 men
5. In people aged 20 and over, low cholesterol is correlated with an increase in death rates
6. Men who stop smoking *and* lower their cholesterol levels have a 2% increase in their death rates
7. 15-year study of 20-95 year olds shows that low cholesterol is related to a higher death rate
8. Older men with lower cholesterol have a 45% higher death rate
9. Men and women with higher cholesterol live longer
10. 18-year study finds that low cholesterol equates with higher death rates in men
11. Low cholesterol contributes to men dying at 42 times the rate of men with higher cholesterol

12 Low cholesterol increases the risk of death by at least 340% in elderly women
13 Death rates increase by 18% for every 1 mmol/L (38 mg/dL) decrease in cholesterol levels
14 Italian study shows that low cholesterol levels lead to a 47% increased mortality rate compared to high cholesterol levels
15 Older men and women with low cholesterol have a 39% higher risk of death
16 20-year study shows those with the lowest cholesterol levels have a 35% increase in death rates compared to those with the highest cholesterol
17 Higher cholesterol levels in both men and women are linked to a longer life
18 People with higher cholesterol levels have a longer lifespan
19 Older people with the highest cholesterol live the longest
20 Low cholesterol levels are an accurate predictor of higher mortality levels in the non-demented elderly
21 High cholesterol levels lead to a longer life in very old people
22 An analysis of 297,574 men and women finds that high cholesterol levels are linked to a longer life
23 Declining cholesterol rates in people over 65 are associated with a 630% increase in death rates
24 30% higher death rate for men with falling cholesterol levels
25 Low cholesterol levels are associated with death in patients admitted to hospital
26 Hospital patients more likely to die with low cholesterol
27 44% increase in death rates for hospital patients with low cholesterol levels

Chapter 2: High levels of both "good" and "bad" cholesterol help you to live longer

28 53-year study shows that low levels of HDL cholesterol lead to an increased death rate
29 High total cholesterol levels and high levels of HDL cholesterol reduce the death rate
30 Eating more saturated fat helps you live to at least 85 years of age
31 High levels of HDL cholesterol are associated with better survival in people aged over 80
32 Low HDL cholesterol levels are associated with increases in deaths from heart disease and cancer
33 Low HDL cholesterol and low LDL cholesterol levels are linked to an earlier death
34 Middle-aged men have lower death rates with higher levels of HDL and LDL cholesterol
35 Low cholesterol levels, and in particular, low levels of LDL cholesterol are associated with a shorter life span
36 High levels of LDL cholesterol lead to a longer life
37 Low LDL cholesterol levels are associated with an earlier death
38 High levels of LDL cholesterol are associated with lower death rates and lower rates of cardiovascular disease

Chapter 3: High cholesterol does not cause cardiovascular disease

39 Higher death rates, increased stroke and heart disease associated with low cholesterol
40 Rates of heart disease deaths are higher with low cholesterol
41 Low cholesterol leads to 30% higher death rates from vascular causes

42 A direct association between falling cholesterol levels and increased death rates from cardiovascular diseases
43 Higher cholesterol levels reduce the risk of cardiovascular diseases
44 Higher cholesterol levels and higher meat consumption are associated with decreased rates of heart disease deaths
45 Heart attack survivors live longer if they have high cholesterol
46 In patients with chronic heart failure, higher cholesterol levels are associated with a longer survival time
47 Death rates in stroke, heart failure, and cancer are elevated in people with low cholesterol levels
48 Heart failure patients with low cholesterol levels have a 240% increased risk of urgent transplant and death
49 Higher cholesterol levels are associated with less severe strokes and lower death rates
50 Low cholesterol levels increase the risk of death from stroke, cancer and all-causes
51 Patients hospitalised with a stroke with low cholesterol have a 117% increased risk of death compared to patients with high cholesterol

Chapter 4: Low cholesterol leads to an early death in many diseases

52 As cholesterol levels are lowered deaths from accidents, suicides and homicides increase by up to 30%
53 Men with the lowest cholesterol have a three-fold increased risk of death from AIDS compared to men with the highest cholesterol

54 Violence, anti-social behaviour and premature death associated with low cholesterol levels
55 Low cholesterol levels predict death in patients with bacteria in the blood
56 Low cholesterol levels are associated with an increase in death rates especially from cancer
57 Men and women with low cholesterol have higher rates of cancer deaths
58 Colon cancer deaths increase in men with low cholesterol
59 Analysis of 519,643 people reveals low cholesterol increases the risk of dying from pancreatic cancer by 27%
60 A rise in total cholesterol reduces the risk of death from cancer and infections in the oldest old
61 A review of 150 studies finds an association between low cholesterol and death from injury
62 Low cholesterol leads to increased rates of deaths from cancer and injuries, and deaths from diseases of the respiratory and digestive systems
63 Low levels of cholesterol are predictive of higher rates of death in patients with end-stage renal disease
64 Kidney failure patients live longer if they have high cholesterol
65 Higher risk of liver diseases with low cholesterol
66 The connection between chronic obstructive pulmonary disease and low cholesterol
67 Low cholesterol levels are associated with higher death rates from respiratory diseases
68 Correlation between low cholesterol levels and rheumatoid arthritis

69 Meningococcal sepsis is associated with low cholesterol levels
70 Both low cholesterol levels and declining cholesterol levels are associated with increased risk of death from suicide in men
71 Those with the lowest cholesterol levels have more than six times the risk of committing suicide than those with the highest cholesterol levels
72 The lower the cholesterol level, the higher the risk of suicide
73 Rising cholesterol levels predict survival, whilst falling cholesterol levels predicts death in people who have suffered severe injuries
74 Men with low cholesterol are susceptible to cancer and tuberculosis

Chapter 5: Low cholesterol leads to an increased prevalence of many diseases

75 After childbirth, women with lower cholesterol levels have major feelings of fatigue and a depressed mood
76 The link between low cholesterol and autism
77 Low cholesterol values may be a risk factor for aggression in children
78 Low cholesterol is significantly associated with schizophrenia
79 People with low cholesterol are more prone to coughs, colds, runny noses and sore throats
80 Crohn's disease and ulcerative colitis link with low cholesterol

81 Low cholesterol levels are associated with higher rates of many infectious diseases
82 Hepatitis C associated with low cholesterol levels
83 High cholesterol *decreases* the risk of intermittent claudication by 10%
84 Men with gout have lower cholesterol than healthy men
85 Poor memory is associated with low cholesterol
86 High cholesterol levels are associated with higher intelligence
87 High levels of cholesterol significantly decrease the risk of Parkinson's
88 High cholesterol levels are associated with a decreased risk of Alzheimer's disease
89 Low cholesterol levels are associated with higher rates of dementia

Chapter 6: The effects of drugs and diet

90 Statins increase the death rate by 150-300%
91 The elderly die earlier when taking statins
92 Men treated with cholesterol lowering drugs have a 142% increase in heart disease deaths
93 The lower your cholesterol is, the more likely you are to die
94 Lower cholesterol levels result in higher cardiac death rates and higher total death rates
95 The adverse health effects of low cholesterol
96 Low fat, low cholesterol diets result in lower cholesterol levels and a higher death rate
97 Low cholesterol, low saturated fat diets lead to a fall in cholesterol levels and an increase in heart attacks and death rates

98 Professor finds the best way to raise HDL (good cholesterol) is to eat saturated fat
99 Lowering cholesterol results in an increase in death rates

Chapter 7: How does cholesterol help us to live longer?

100 Cholesterol supplementation benefits patients with Smith-Lemli-Opitz syndrome
101 High cholesterol levels strengthen the immune system

Index

(Refers to paper numbers)

abscess of the anal and rectal region 81
abscess of the intestine 81
abscess on the brain 81
abscess on the spinal cord 81
accidents 52
adrenal failure 95
adrenal hormones 100
African-Americans 35
Afsarmanesh, Dr Nasim 48
aggression 76, 77
AIDS 53
Akerblom, JL 35
albumin 29
aldactone 36
aldosterone 100
androstenedione 100
Alzheimer's 88
Anderson, KM 42
Aneja, Dr Alka 76
Ansary-Moghaddam, Dr Alireza 59
Anti-social personality disorder (ASPD) 54
appendicitis 81

arthropathy 81
asbestosis 67
asthma 67
Atmaca, M 78
Atrens, Dr Dale 99
autism 76

bacteremia 55, 81
bartholin cyst 81
beef 29
Behar, Professor Solomon 40
behaviour 100
beta blockers 92
bile acid 100
boils 81
Bowden, RG 63
Bradford, Dr Reagan 90
brain 100
Brescianini, Dr Sonia 14
bronchitis 67

cancer 40, 47, 50, 56, 57, 60, 62, 74, 95, 101
cancer, colon, 58
cancer, pancreatic 59

carbuncle 81
cardiac death 94
cardiovascular disease 42, 43, 99
Casiglia, Dr Edoardo 17
CD4+ cells 101
CD8+ cells 101
cell membranes 100
cellulitis 81
cervicitis 81
Chen, Professor Zhengming 3, 65
chicken 29
childbirth 75
chlamydia 81
cholangitis 81
cholesterol, dietary
    aggression 76
    autism 76
    behaviour 100
    gastrointestinal 100
    growth 76
    hyperactivity 76
    infections 76, 100
    irritability 76
    photosensitivity 100
    puberty 100
    self-injury 76
    skin rash 100
    sleep 76
    Smith-Lemli-Opitz syndrome 76
    social interactions 76
    tantrums 76
    trichotillomania 76
cholesterol functions
    adrenal hormones 100
    aldosterone 100
    androstenedione 100
    bile acid 100
    brain 100
    CD4+ cells 101
    CD8+ cells 101
    cell membranes 100
    corticosterone 100
    cortisol 100
    cytokines 101
    embryonic development 100
    estrogen 100
    hormones 100
    interleukin-2 101
    lymphocytes 101
    myelin 100
    nerve cells 100
    progesterone 100
    repairs damaged arteries 100
    synapses 100
    T-cells 101
    testosterone 100
    vitamin A 100
    vitamin D 100
    vitamin E 100
    vitamin K 100
cholesterol levels
    abscess of the anal and rectal region 81
    abscess of the intestine 81

abscess on the brain 81
abscess on the spinal cord 81
accidents 52
adrenal failure 95
adrenal hormones 100
aggression 77
AIDS 53
aldosterone 100
Alzheimer's 88
androstenedione 100
anti-social personality disorder (ASPD) 54
appendicitis 81
arthropathy 81
asbestosis 67
asthma 67
autism 76
bacteremia 55, 81
bartholin cyst 81
behaviour 100
bile acid 100
boils 81
brain 100
bronchitis 67
cancer 40, 47, 50, 56, 57, 60, 62, 74, 95, 101
cancer colon 58
cancer, pancreatic 59
carbuncle 81
cardiac death 94
cardiovascular disease 42, 43, 99
CD4+ cells 101
CD8+ cells 101
cell membranes 100
cellulitis 81
cervicitis 81
childbirth 75
chlamydia 81
cholangitis 81
chronic obstructive pulmonary disease 66
cognition 86
colds 79
conduct disorder 54
corticosterone 100
cortisol 100
coughs 79
crohn's 80
cytokines 101
death rates 1, 2, 3, 4, 5, 6, 7, 8, 9, 10, 11, 12, 13, 14, 15, 16, 17, 18, 19, 20, 21, 22, 23, 24, 25, 26, 27, 29, 32, 33, 34, 35, 39, 40, 42, 43, 47, 49, 50, 52, 56, 57, 60, 74, 91, 92, 93, 94, 95, 96, 97, 99
death rates hospital patients 25, 26, 27
death rates men, 2, 4, 6, 8, 10, 11, 16, 24, 50, 52, 56, 74
death rates elderly 1, 8, 11, 12, 13, 14, 15, 16, 17, 18, 19, 20, 21, 23, 27, 29, 33, 39, 60, 91
death rates non heart disease 40

death rates Pima Indians 5
death rates women 12, 20
dementia 89
depression 75
digestive system 62
dilated cardiomyopathy 48
diverticulosis 81
eczema 81
embryonic development 100
emphysema 67
empyema 67
encephalitis 81
endocarditis 81
endotoxic shock 81
estrogen 100
fatigue 75
gallbladder 81
gangrene 81
gastrointestinal 100
genito-urinary infections 81
gonorrhoea 81
gout 84
growth 76
heart attack 45, 97
heart disease 39, 40, 41, 44, 92, 96
heart failure 41, 46, 47, 48
heart transplant 48
hemodynamics 48
hemorrhagic stroke 47, 50
hepatitis B 65
hepatitis C 82
herpes 81

homicides 52
hormones 100
hyperactivity 76
immune system 101
impetigo 81
infections 60, 76, 95, 100
infective myositis 81
inflammatory bowel disease 80
influenza 67
injury 61, 62, 73
interleulin-2 101
intermittent claudication 83
intestinal infections 81
intracerebral hemorrhage 95
irritability 76
ischemic heart disease 66
ischemic stroke 51
kidney disease 63, 64, 81
Iso, H 57
laryngitis 67
left ventricular ejection fraction 48
liver disease 65, 81
lung disease 67
lymphadenitis 81
lymphocytes 101
mediastinitis 67
memory 85
meningitis 81
meningococcal sepsis 69
mental disorder 95
minor illnesses 79

muscle-skeletal infections 81
myelin 100
myelitis 81
oophoritis 81
nerve cells 100
New York Heart Association (NYHA) Functional Classification 48
non-cancer 62
non-cardiovascular 62
non-heart disease 40
nonischemic systolic heart failure 48
non-vascular disease 41
osteomyelitis 81
Parkinson's 87
pelvic inflammatory disease 81
periostitis 81
physicians 95
photosensitivity 100
pilonidal cyst 81
pleurisy 67
pneumonia 67
pneumuconiosis 67
poor health 39, 56
progesterone 100
puberty 100
pulmonary congestion 67
pulmonary fibrosis 67
pyoderma 81
respiratory system 62, 67, 74
rhinitis 67

rheumatic pneumonia 67
rheumatoid arthritis 68
ringworm 81
rotavirus 81
runny nose 79
salmonella 81
salpingitis 81
Scandinavian Stroke Scale 49
schizophrenia 78
self-injury 76
septicaemia 81
septic shock 81
sinusitis 67
skin rash 100
sleep 76
Smith-Lemli-Opitz syndrome 76, 99
social interaction 76
sore throat 79
stroke 39, 41, 49, 66
sudden cardiac death 66
sudden death 97
suicide 52, 70, 71, 72, 78, 95
synapses 100
syphilis 81
tantrums 76
T-cells 101
testosterone 100
thrush 81
tonsillitis 67
trauma 73
trichomoniasis 81

trichotillomania 76
tuberculosis 74
ulcerative colitis 80
urinary tract infections 81
vaginitis 81
vascular disease 41
venereal diseases 81
violence 54, 78, 92
viral hepatitis 81
vitamin A 100
vitamin D 100
vitamin E 100
vitamin K 100
cholesterol lowering drug therapy 93
chronic obstructive pulmonary disease 66
clofibrate 92
cognition 86
colds 79
conduct disorder 54
corticosterone 100
cortisol 100
coughs 79
Cowan, Dr Linda 58
Crohn's 80
Crook, MA 26
Cullen, P 34
Cummings, Professor Peter 61
cytokines 101

demadex 36
dementia 89
depression 75
digestive system 62
dilated cardiomyopathy 48
diuretics 36, 92
diuril 36
diverticulosis 81
Dontas, Professor Anastasios 21
Dunham, Dr Michael 73

eczema 81
eggs 29
Elias, Dr Penelope 86
Elias, ER 100
Ellison, L 71
embryonic development 100
emphysema 67
empyema 67
encephalitis 81
endocarditis 81
endotoxic shock 81
enduron 36
esidrix 36
estrogen 100

Fagot-Campagna, Dr Anne 5
fat, dietary 98, 99
fatigue 75,
fava beans 101
fish 29
Foody, Dr Joanne 45
Forette, Dr Bernard 12
Frantz, ID 97

Frick, Dr M Hekki 94
Fried, Dr Linda 36

gallbladder 81
gangrene 81
gastrointestinal 100
gemfibrozil
    cardiac deaths 94
    death rates 94
genito-urinary infections 81
gonorrhoea 81
gout 84
Grant, Dr Mark 23
growth 76

Harris, Dr Tamara 1
HDL cholesterol 29
    Alzheimer's 88
    cancer 32
    colds 79
    coughs 79
    Crohn's 80
    death rates 28, 30, 31, 32, 33, 34, 35
    dementia 89
    gout 84
    heart disease 28, 32
    hepatitis C 82
    inflammatory bowel disease 80
    memory 85
    minor illnesses 79
    runny nose 79
    saturated fat 28, 29, 30, 31, 32, 98
    sore throat 79
    ulcerative colitis 80
HDL 2 cholesterol 28
HDL 3 cholesterol 28
heart attack 45, 97
heart disease 39, 40, 41, 44, 92, 99
heart failure 41, 46, 47, 48
heart transplant 48
Hematocrit 11, 21
hemodialysis 64
hemodynamics 48
hemorrhagic stroke 47, 50
hepatitis B 65
hepatitis C 82
herpes 81
Hispanics 35
homicides 52
hormones 100
Hu, P 15
Hulley, Dr Stephen 62
Hurt-Camejo, Professor Eva 68
hydrozyne 36
Hyman, DJ 79
hyperactivity 76

immune system 101
impetigo 81
infections 60, 76, 95, 100
infective myositis 81
inflammatory bowel disease 80

influenza 67
injury 61, 62, 73
interleukin-2 101
intermittent claudication 83
intestinal infections 81
intracerebral hemorrhage 95
Iribarren, Dr Carlos 24, 50, 67, 81
irritability 76
ischemic heart disease 66
ischemic stroke 51
Iseki, K 64
Isles, Dr Christopher 9
Ives, Diane 18

Jacobelli, S 84
Jacobs, D 22
Jonsson, A 19

Kame, C 93
Kidney beans 101
kidney disease 63, 64, 81
Kozarevic, DJ 74

Landi, Dr Francesco 31
laryngitis 67
lasix 36
LDL cholesterol
    Alzheimer's 88
    cancer, colon 58
    death rates 33, 34, 35, 36, 37, 38
    gout 84
    heart disease 38, 44
    hepatitis C 82
    kidney disease 63
    suicide 72
Lee, HJ 72
left ventricular ejection fraction 48
liver 11
liver disease 65, 81, 90
Lonneke, M 87
Low-cholesterol diets
    death rate 96, 97
    heart attack 97
    heart disease 96
    sudden death 97
low-fat diets
    cardiovascular disease 99
    death rate 96
    heart disease 96
low saturated-fat diets
    death rates 97
    heart attacks 97
    sudden death 97
lozol 36
leguminous beans 101
lung disese 67
Luoma, Professor Pauli 44
lymphadenitis 81
lymphocytes 101

meat 11, 44
mediastinitis 67
memory 85

meningitis 81
meningococcal sepsis 69
Mensink, Professor Ronald 98
mental disorders 95
minor illnesses 79
Muldoon, MF 52, 101
muscle damage 90
muscle-skeletal infections 81
myelin 100
myelitis 81

Nago, Dr Naoki 47
Nasta, MT 75
Neaton, Professor James 4, 53
nerve cells 100
New York Heart Association (NYHA) Functional Classification 48
non heart disease 40
nonischemic systolic heart failure 48
nonvascular disease 41

Ogushi, Y 37
Olsen TS 49
Onder, Dr Graziano 27
oophoritis 81
oretic 36
osteomyelitis 81

Parkinson's 87
Passons VM 83
pelvic inflammatory disease 81

periostitis 81
Petersen, Dr Line Kirkeby 91
Petursson, H 43
Photosensitivity 100
physicians 95
phytohemagglutinin 101
pilonidal cyst 81
Pima Indians 5
pleurisy 67
pneumonia 67
pneumuconiosis 67
poor health 39
pork 29
probucol 92
progesterone 100
puberty 100
pulmonary congestion 67
pulmonary fibrosis 67
pyoderma 81

Rahilly-Tierney, CR 30
Raiha, Dr Ismo 41
Rauchhaus, Dr Mathias 46
red kidney beans 101
red meat 11
Reitz, Dr Christiane 88
Repo-Tiihonen, Dr Eila 54
respiratory system 62, 67, 74
reverse epidemiology 63
rhinitis 67
rheumatic pneumonia 67
rheumatoid arthritis 68
Richardson, JP 55

ringworm 81
Ripollés Piquer B 80
rotavirus 81
Rudman, Dr Daniel 8, 11
runny nose 79

salmonella 81
salpingitis 81
saluron 36
saturated fat 28, 29, 30, 31, 32, 98
Scandinavian Stroke Scale 49
Schatz, Professor Irwin 16
schizophrenia 78
Schupf, Dr Nicole 20
self-injury 76
septic shock 81
septicaemia 81
Siagris, Dr Dimitrios 82
Sin, Dr Don 66
Singh-Manoux, Dr Archana 85
sinusitis 67
skin rash 100
sleep 76
Smith-Lemli-Opitz syndrome 76, 100
smoking 6
social interaction 76
Song, Dr Jun-xian 95
sore throat 79
statins 95
    death rates 90, 91
    elderly 91
    liver disease 90
    muscle damage 90
    transaminase levels 90
Stemmermann, GN 10
Strandberg, TE 92
stroke 39, 41, 49, 66
sudden cardiac death 66
sudden death 97
suicide 52, 70, 71, 72, 78, 95
synapses 100
syphilis 81

tantrums 76
T-cells 101
testosterone 100
thalitone 36
thrush 81
Tikhonoff, Dr Valerie 38
Tilvis, Professor Reijo 39
tonsillitis 67
transaminase levels 90
trauma 73
trichomoniasis 81
trichotillomania 76
triglycerides 98
tuberculosis 74
Tuikkala P 13

ulcerative colitis 80
Ulmer, H 7
urinary tract infections 81

vaginitis 81
vascular disease 41

venereal diseases 81
Vermont CL 69
violence 54, 78, 92
viral hepatitis 81
vitamin A 100
vitamin D 100
vitamin E 44, 100
vitamin K 100
Volpato, Dr Stefano 29

Wannamethee, Dr Goya 56
water pills 36

Weverling-Rijnsburger, AW 33, 60
Williams, Paul 28
Wilson, Professor Peter 32
Windler, Professor Eberhard 25
Woodhill, JM 96

Yun-Mi Song 2

zaroxolyn 36
Zhang, Professor Jian 77
Zuliani, G 51, 89
Zureik, Dr Mahmoud 70